Surprises in Theoretical Physics

PRINCETON SERIES IN PHYSICS

Edited by Phillip W. Anderson, Arther S. Wightman, and Sam B. Treiman (published since 1976)

Surprises in Theoretical Physics

by
Rudolf Peierls

Princeton Series in Physics

Princeton University Press
Princeton, New Jersey 1979

Library of Congress Cataloging in Publication Data

Peierls. Rudolf Ernst, Sir, 1907–
Surprises in theoretical physics.

(Princeton series in physics)
Based on lectures given at the University of Washington in spring 1977, and at
the Institut de Physique Nucleaire, Université de Paris-Sud, Orsay, in winter 1977–78.

1. Mathematical physics. I. Title.
QC20.P35   530.1   79-84009
ISBN 0-691-08241-3
ISBN 0-691-08242-1 pbk.

CONTENTS

PREFACE

PREFACE

This book is based on lectures given in the University of Washington in the Spring of 1977, and in the Institut de Physique Nucléaire, Université de Paris-Sud, Orsay, during the Winter 1977–78. The purpose of the lectures was to present a number of examples in which a plausible expectation is not borne out by a more careful analysis. In some cases the outcome of a calculation is contrary to what our physical intuition appears to demand. In other cases an approximation which looks convincing turns out to be unjustified, or one that looks unreasonable turns out to be adequate.

My intention was not, however, to give the impression that theoretical physics is a hazardous game in which we can never foresee what surprises the detailed calculation will reveal. On the contrary, all the surprises I discuss have rational explanations, and these are mostly very simple, at least in principle. In other words, we should not have been surprised if only we had thought sufficiently deeply about our problem in advance. The study of such surprises can therefore help us to avoid errors in future problems, by showing us what kind of possibilities to look for.

The selection of examples is very subjective. Most concern surprises experienced by myself or my collaborators, others I found fascinating when I heard of them. No doubt the list could be extended greatly, but any attempt at completeness would defeat the object of the lectures, and of this little book, which is to illustrate, and to entertain.

The examples are taken from many branches of physical theory, mostly within quantum physics, but I have tried to give in each case enough background to make the presentation intelligible without specialist knowledge. They are grouped roughly according to the fields of physics to which they relate, but there is otherwise little relation or similarity between the items in each group.

The idea of publishing the substance of the lectures as a book arose from their kind reception by their audiences, which comprised young and old theoreticians as well as experimental physicists interested in theory. Many of them, both at Seattle and at Orsay, asked pertinent questions, or raised objections, which helped to clarify important points. I recall particularly constructive comments by N. Austern, E. Domany, B. Jancovici, and C. Marty. I am also obliged to Norman Austern for helpful criticism of the written text. During most of the writing I enjoyed the hospitality of the Theoretical Physics Division of the Institut de Physique Nucléaire at Orsay.

Surprises in Theoretical Physics

1. GENERAL QUANTUM MECHANICS

1.1 BORN APPROXIMATION FOR SHORT-RANGE INTERACTION

Born approximation is a familiar and convenient approximation for handling scattering problems. It is adequate, or at least informative, in so many cases that we tend to develop the habit of using its first-order term without always checking the conditions for its applicability.

To be sure, we are likely to stop and take notice if the application of first-order Born approximation results in a large phase shift, since a large phase shift cannot very well be caused by a weak perturbing potential. But one is tempted to trust the approximation if it predicts a small phase shift, or if we know that the true phase shift is small. Neither of these statements is, however, sufficient to ensure the validity of first-order Born approximation, and this is particularly important in the case of short-range forces.

In order to understand the situation we recall that the scattering amplitude in Born approximation is, to first order, given by

$$A_B = - \frac{m}{2\pi\hbar^2} \langle \phi_f | U | \phi_i \rangle, \qquad (1.1.1)$$

where m is the mass of the scattered particle, U the scattering potential, and ϕ_i, ϕ_f are plane waves for the incident and final states. The exact scattering amplitude, on the other hand, is

$$A = - \frac{m}{2\pi\hbar^2} \langle \phi_f | U | \psi_i \rangle, \qquad (1.1.2)$$

where ψ_i is the complete solution of the Schrödinger equation in the given potential, and contains both the incident wave and the scattered waves. One is now tempted to reason that, if the scattering is weak, the difference between ϕ_i and ψ_i should be small, and (1.1.1) a reasonable approximation to (1.1.2). This reasoning is unsafe, particularly for a short-range potential, since the approximation depends on ψ_i being similar to ϕ_i in the region in which the potential acts.

If the range of the potential is short compared to the wavelength, we are primarily concerned with s-wave scattering. Then we can use instead of the above relations the expressions

$$k \sin \delta_B = -\frac{m}{\hbar^2} \int_0^\infty \phi_0^2 U \, dr \qquad (1.1.3)$$

and

$$k \sin \delta = -\frac{m}{\hbar^2} \int_0^\infty \phi_0 \psi_0 U \, dr, \qquad (1.1.4)$$

where δ is the phase shift and δ_B its value to first order in Born approximation, and ϕ_0, ψ_0 are the radial wave functions (including the usual factor r). Figure 1.1 shows these functions qualitatively for a short-range repulsive potential. In the case illustrated, δ is small, and ψ_0 is very little different from ϕ_0 almost everywhere, except for small r, where both are small, but where ψ_o is much smaller than ϕ_0; this is just the region where U is appreciable, and that is the region which matters for the integral (1.1.4).

Consider, for example, a square barrier potential of height B and radius a. The s-wave phase shift is then given by the equation

$$\frac{\tan(ka + \delta)}{ka} = \frac{\tan Ka}{Ka}, \qquad (1.1.5)$$

where k is the wave number of the scattered particle, and

$$K^2 = k^2 - \frac{2m}{\hbar^2} B. \qquad (1.1.6)$$

K may well be imaginary. If the range a is much less than the wave-

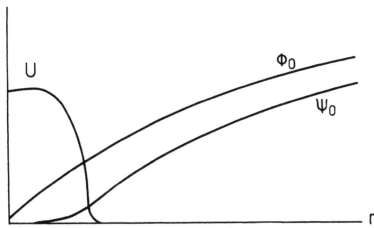

Figure 1.1   Exact (ψ_0) and free-particle (ϕ_0) radial s-wave functions in a short-range repulsive potential.

length, $ka \ll 1$. In that case δ is also small, regardless of the magnitude of U, and we may approximate (1.1.5) by

$$\delta = ka \left(\frac{\tan Ka}{Ka} - 1 \right). \qquad (1.1.7)$$

If Ka is also small, this reduces to

$$\delta = -\frac{2m}{\hbar^2} \frac{ka}{3} Ba^2 = \delta_B, \qquad (1.1.8)$$

which is also the expression for the phase shift in Born approximation.

This result is valid if

$$\frac{2m}{\hbar^2} Ba^2 \ll 1, \qquad (1.1.9)$$

and then the expression (1.1.8) is small a fortiori.

However, if B is greater than $\hbar^2/2ma^2$, but less than $\hbar^2/2mka^3$, the Born phase shift (1.1.8) is still small, but no longer gives the

correct answer. This therefore is a situation in which both the true
phase shift and that calculated in first-order Born approximation
are small, yet the approximation is invalid.

Alternatively, if $B > \hbar^2/2mka^2$, the Born phase shift becomes
large, while the true phase shift remains small.

This situation is worth remembering, to avoid surprises in
scattering calculations.

1.2. SHADOW SCATTERING

A very general identity governing all scattering processes is the
so-called "optical theorem", a special case of the even more general
unitarity relation, which ensures the conservation of probability.
For a spinless particle the optical theorem takes the form

$$\frac{4\pi}{k} \, Im A(0) = \sigma_{tot}, \qquad (1.2.1)$$

where k is the wave number, $A(\theta)$ the elastic scattering amplitude
at angle θ, and σ_{tot} the total cross section. A related identity holds
for the scattering amplitude and total cross section in the ℓth partial wave:

$$\sigma_{tot}^{\ell} = \frac{4\pi}{k} \, Im A_{\ell}, \qquad (1.2.2)$$

where σ_{tot}^{ℓ} is the total scattering cross section in the partial wave
of angular momentum $\ell\hbar$, and A_{ℓ} the partial elastic amplitude.
The total contribution of the partial wave to the elastic scattering is

$$\sigma_{el}^{\ell} = \frac{4\pi}{2\ell + 1} |A_{\ell}|^2. \qquad (1.2.3)$$

The square of the imaginary part of a quantity is less than its square
modulus, and we find the inequality

$$(\sigma_{tot}^{\ell})^2 \leq q_{\ell}\sigma_{el}^{\ell}; \quad q_{\ell} = \frac{4\pi(2\ell + 1)}{_{\ell}k^2}. \qquad (1.2.4)$$

The quantity q_ℓ introduced here is a measure of the total intensity incident in the ℓth partial wave; it is natural that it should appear in the conservation laws.

Now the inelastic scattering is the difference between the total and the elastic part, so that the inequality (1.2.4) implies

$$\sigma_{inel}^\ell \leq \sqrt{(q\ \sigma_{el}^\ell)} - \sigma_{el}^\ell. \tag{1.2.5}$$

The right-hand side has a maximum when $\sigma_{el}^\ell = \tfrac{1}{4}q_\ell$, and is then equal to $\tfrac{1}{4}q_\ell$. It follows therefore that the inelastic scattering cross section in the ℓth partial wave cannot exceed $\tfrac{1}{4}q_\ell$, and it cannot reach this value unless there is an equally large elastic contribution.

Let us apply this to the case of a large black sphere, with a radius R which is very large compared to the wavelength. In that case the ℓth partial wave corresponds, to a good approximation, to a classical motion with an impact parameter ℓ/k. In this classical limit we expect, therefore, that for all partial waves with $\ell < kR$, corresponding to motions in which the particle hits the sphere, there will be maximum absorption, i.e., maximum inelasticity, and no absorption for $\ell > kR$. Because of quantum effects the transition will not, in fact, be discontinuous, but will be spread over a few values of ℓ. If kR is large enough this is unimportant. We conclude that there should be an inelastic cross section of

$$\sum_{\ell=0}^{kR} \frac{1}{4} q_\ell = \frac{\pi}{k^2} \sum_{\ell=0}^{kR} (2\ell + 1) = \pi R^2, \tag{1.2.6}$$

which seems reasonable enough, since this is the geometrical cross section of the sphere. However, our argument allowed us to conclude that this requires an equal elastic contribution.

This is, at first sight, a very surprising conclusion. In the collision between two classical objects like billiard balls or snowballs (even the best billiard ball will dissipate *some* energy in friction, so that the collision is never elastic in the sense of the strict definition) it is most surprising to hear that there should be an elastic cross section equal to the geometric one in addition to the expected inelastic one.

To understand this, we note, first of all, that the predicted elastic scattering is concentrated into a small solid angle around the forward direction. The elastic amplitude at angle θ is

$$A(\theta) = \sum_{\ell} A_{\ell} P_{\ell} (\cos \theta). \tag{1.2.7}$$

From the discussion above, we know that the first kR terms will have values of A_{ℓ} which are purely imaginary, with positive imaginary parts. Therefore all terms contributing to $A(\theta)$ are in phase and add. However, for non-zero values of θ the Legendre polynomials begin to oscillate. The angle at which P_{ℓ} has its first zero is of the order of $1/\ell$. For angles larger than $1/kR$ the sum (1.2.7) will contain many contributions of opposite sign, and we can expect it to be negligible. (This can indeed be proved.) In these circumstances the elastic scattering can be observed only at angles of the order of $1/kR = \lambda/2\pi R$, where λ is the wavelength, and any rough estimate for a billiard ball or other macroscopic object shows that such an observation would not be feasible in practice.

The physical origin of the elastic scattering can be made clear by the following consideration: In the case of a wave incident on a large absorbing sphere, as in Figure 1.2, there must be a shadow behind the sphere, of a cross section equal to the geometrical cross section of the sphere, and extending downstream to a distance beyond which diffraction effects fill in the shadow. But in scattering theory we write the complete wave function for the elastic channel as

$$\psi = \phi_i + \psi_1, \tag{1.2.8}$$

where ϕ_i is again the incident wave and ψ_1 the scattered wave. Behind the sphere, in the shadow, the wave function must vanish. This requires that the second term of (1.2.8) be opposite and equal to the first. This holds over the width of the shadow, and the flux contained in ψ_1 is therefore equal to the flux which the incident wave would have carried through that area. This explains why the

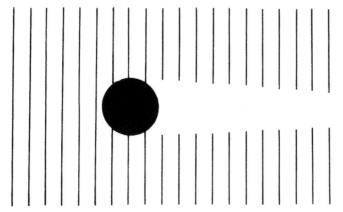

Figure 1.2   Shadow scattering.

flux associated with the wave function ψ_1 equals the amount incident on an area equal to the geometrical cross section.

However, starting from the classical situation, we still do not find it easy to accept that the shadow should amount to actual scattering. It follows from general scattering theory that anything other than the incident wave forms part of the scattered wave, and contributes an asymptotically outgoing wave at large distances.

It is, however, important to remember that, in the present situation, one may have to go to very large distances before the asymptotic behavior is reached. We had seen above that the scattering angles in the elastic wave are only of the order of $1/kR$. This means that the asymptotic parts of ψ_1 spread only through such small angles, and since they are all coherent initially over a width R, they cannot have reached their asymptotic form in a distance less than kR^2, or about R^2/λ. If we wanted to verify the predicted scattering in the case of a billiard ball, we would therefore not only require an instrument which can resolve extremely small angles of deflection, but we would have to carry out the observation at distances greater than astronomical.

Having recovered from the initial surprise, we note that the shadow scattering can have quite practical implications in less extreme circumstances. If particles of high momentum are scattered, for example, by nuclei, one very easily meets cases in which kR is large, so that our discussion applies. However, the values of kR are then not unreasonably large, and the scattering angles not prohibitively small. In addition, the distance at which the observation is carried out is then normally much greater than $R^2 k$, so that the phenomenon is always seen as small-angle scattering, and not as a shadow.

Historical note: The optical theorem and shadow scattering were first noticed by the author in connection with work for a joint paper with Niels Bohr and G. Placzek, which never progressed beyond the draft stage, but has been cited frequently.

1.3. WAVES AND PARTICLES

One of the most basic ideas of quantum mechanics is the analogy between light and matter, first recognized by L. de Broglie. This teaches us that both possess wave and particle aspects, and we have all met, at the beginning of our acquaintance with quantum mechanics, the manifestations of the particle aspect of light in Planck's radiation formula or in the Compton effect, and the wave nature of electrons in diffraction experiments.

From this, it might appear an accident of history that physicists originally encountered only the wave aspects of light and only the corpuscular aspects of particles with mass, such as electrons. It therefore comes as a surprise to discover that this is no accident at all, and that the analogy between light and matter has very severe limitations. Indeed, as we shall see, there can be no classical field theory for electrons, and no classical particle dynamics for photons.

To see this, we note that a field is characterized by its amplitude and phase. For a classical treatment to be applicable, both must be specified with a negligible uncertainty. But there is an uncertainty relation between amplitude and phase. If, for simplicity, we consider a single scalar field, we can write the wave as

$$\phi = A e^{i\gamma}, \tag{1.3.1}$$

where γ is a phase, and A is normalized on such a scale that

$$A^2 = N \tag{1.3.2}$$

is the number of quanta per mode. Then the uncertainty relation is, adequately for our purpose (a refinement will be discussed in section 1.4),

$$\delta N \, \delta \gamma \geq 1. \tag{1.3.3}$$

This shows that any knowledge of γ requires an uncertainty in N. Hence any device capable of measuring the field, including its phase, must be capable of altering the number of quanta by an indeterminate amount. This is impossible for an electron, both because of its charge and because of its half-integral spin. Electrons cannot be created or destroyed singly. An observation associated with the creation or destruction of an electron pair can provide information only about a pair wave function, or, in other words, about the product of an electron and a positron field.

A further difficulty applies to all fermions, whether charged or neutral, if we want not merely to have some information about the phase, but want to treat the field as a classical quantity. Then the uncertainty in the phase should be small compared to unity, and the uncertainty in N should be small compared to N itself. But from (1.3.3) this requires that

$$N \gg 1. \tag{1.3.4}$$

If we remember that N is the number per mode, we see that this requirement contradicts the Pauli principle, which allows at most one particle per mode.

We conclude that there can be no classical field description for fermions. The difficulty does not apply to a boson field. For radiation it is indeed true that the typical devices for detecting electromagnetic fields depend on a coupling term of the form $j_\mu A_\mu$ in the Hamiltonian, where j is the current density, and the vector potential A contains photon creation and annihilation operators.

The requirement of large N is also readily satisfied when we are dealing with classical electromagnetic fields. In the case of a continuous spectrum it is easy to see that N should be interpreted, in order of magnitude, as the number of photons in a volume λ^3, where λ is the wavelength. For a $1\,kW$ transmitter at a frequency of 1 megahertz, one can estimate that N drops to about 1 only at a distance exceeding 10^{10} km. Similarly, one can verify that a laser beam is intense enough to allow amplitude and phase to be known with good precision.

For neutral bosons, for example for neutral pions, the difficulties connected with conservation of their number, or from the Pauli principle, do not apply. However, since they have a non-vanishing rest mass, changing their number involves a substantial change in energy. Superpositions of states with different particle numbers must therefore show very rapid oscillations in time, at least for free fields. While knowledge of the phase of the field is then not impossible in principle, it is then only of very academic interest.

The other side of the problem relates to the possibility of treating photons as classical particles. This would require treating the position of a photon as an observable. If this were possible, quantum mechanics would have to be able to predict the outcome of such observations. There should therefore exist an expression for the probability of finding the photon in a given volume element. Such an expression would have to behave like a density, i.e., it should be the time component of a four-vector. Being a probability, it should also be positive definite, i.e., it should equal a square or a sum of squares of field quantities. But in the case of the electromagnetic field, the field quantities are vectors (like the vector potential) and second-rank tensors (like the field amplitudes). All are tensors of integral rank, and the same is true of their derivatives. A square of a component of any tensor of integral rank is necessarily part of a tensor of even rank, and cannot belong to a four-vector.

The above argument was used by Dirac in his classical paper on the relativistic wave equation for the electron, and led him to attribute to the electron a spinor as wave function, which amounts to a tensor of half-integral rank. Although the answer for the electron was right, we learned later from the work of Pauli and Weisskopf that the postulate of the existence of an observable position was not self-evident, otherwise we could not believe in the existence of pions and other bosons.

The conclusion that the position of a photon cannot be observable is somewhat surprising. One must remember, of course, that we can certainly localize photons to the accuracy of geometrical optics, within which particle and wave description are equivalent. We are talking about the determination of a position within the wave packet or beam of geometrical optics, analogous to an observation which might locate an electron within the hydrogen atom.

It might be objected that there exist experiments which give the photon distribution in space, such as the Lippman fringes seen in a photographic plate in the standing wave formed at a reflecting surface. However, the photographic plate responds to the electric vector in the light wave. If we could make a detection device based on a magnetic dipole transition, it would respond to the magnetic vector, and in the Lippman experiment would show maxima where the conventional photographic plate has minima, and vice versa. A moving photographic plate would respond to the electric vector in its own rest frame, which is a linear combination of the electric and magnetic vectors in the laboratory frame, so the fringes are shifted. (This is closely connected with the fact that we could not find a density with the correct relativistic transformation property.)

The same difficulty should apply to all bosons, which have integral spin, and therefore fields which are tensors of integral rank. How can we then understand why the use of a Schrödinger equation for π-mesic atoms seems to give quite satisfactory results, with no apparent difficulty of interpretation? The explanation is that, for a meson field, there exists a positive definite expression for the energy density. In the non-relativistic limit, the kinetic and potential energies of the meson are negligible compared to its rest energy; one may thus assume the energy per particle to be mc^2, and therefore take the particle density as $1/mc^2$ times the energy density.

Thus the position of a boson can be given an approximate meaning in the non-relativistic domain. For the electron the position is actually no better, if we remember that the Dirac equation admits solutions with negative energy, which have to be re-interpreted in the light of positron theory. In consequence, any observation of position in which the momentum uncertainty becomes relativistic is liable to lead to pair creation, and therefore removes us from the one-electron picture within which the concept of "the position of the electron" is applicable.

We see, therefore, that ultimately the concept of a position is applicable to both fermions and bosons in the non-relativistic region, and to neither of them beyond. The exceptional role of the photon is due to its vanishing mass, by which it can never be non-relativistic. It is just this vanishing mass which makes it easy to create and absorb photons, and therefore helps to realize the limit of a classical field.

Historical note: The considerations summarized above were first discussed by Pauli and the author (see W. Pauli, Z. *Physik*, 80, 573, 1933). A brief summary can also be found in *Reports on Progress in Physics*, 18, 471, 1955.

1.4.   ANGLE OPERATOR

For some problems in elementary quantum mechanics it is convenient to introduce plane polar coordinates:

$$x = r \cos \theta$$
$$y = r \sin \theta \tag{1.4.1}$$

Then the angle, θ, becomes one of the dynamic variables, and it seems natural to associate with it an operator, as for other observables.

The conjugate variable to θ is then the angular momentum, M.

If this is measured in units of $\hbar$, it is easy to show that M can be represented by the operator

$$M = -i \frac{\partial}{\partial \theta}. \tag{1.4.2}$$

The commutator is then

$$[M, \theta] = -i. \tag{1.4.3}$$

(We have, in fact, used a very similar relation in deriving (1.3.3), and this is therefore the right place to look at the possible complications.)

We also know that M has integral eigenvalues. In a representation in which M is diagonal, (1.4.3) reads

$$(M - M')\langle M|\theta|M'\rangle = -i\delta_{M, M'}. \tag{1.4.4}$$

But this is an impossible relation, since for $M = M'$ the left-hand side vanishes, but the right-hand side does not.

Similarly, if we introduce the Hamiltonian for free rotation by

$$H = \frac{\hbar^2}{2I} M^2, \tag{1.4.5}$$

where I is the moment of inertia, we find

$$\frac{\partial \theta}{\partial t} = -\frac{i}{\hbar} [H, \theta] = \frac{\hbar}{I} M, \tag{1.4.6}$$

which is sensible, but also, again in the M representation,

$$\frac{i}{\hbar} (E_M - E_{M'})\langle M|\theta|M'\rangle = \frac{\hbar}{I}\delta_{M,M'}M, \tag{1.4.7}$$

which is not.

The resolution of the paradox lies in the fact that θ is not an observable. The definition (1.4.1) contains only periodic functions of θ, and therefore does not allow us to distinguish values of θ which differ by a multiple of 2π. The difficulty is not resolved by a convention which would restrict the values of θ to a given interval, say between $-\pi$ and $+\pi$, because, while this would make the definition single-valued, it also makes it discontinuous, and this is inconsistent with relations such as (1.4.2).

While θ thus cannot be treated as an observable, we can introduce

$$\xi = e^{i\theta} \tag{1.4.8}$$

instead. ξ is a single-valued function of the Cartesian coordinates,

$$\xi = \frac{x + iy}{\sqrt{(x^2 + y^2)}}. \tag{1.4.9}$$

Admittedly ξ is not Hermitian, but it is unitary, and therefore has perfectly good eigenvalues. (The fainthearted can use the real and imaginary parts of ξ separately.)

Any physically meaningful statement involving θ can also be expressed in terms of ξ. The commutation law becomes, in place of (1.4.3):

$$[M, \xi] = -i\frac{\partial \xi}{\partial \theta} = \xi, \tag{1.4.10}$$

and in the M representation

$$(M - M')\langle M|\xi|M'\rangle = \langle M|\xi|M'\rangle, \tag{1.4.11}$$

which tells us that ξ is a raising operator, connecting only states for which $M - M' = 1$.

An example of physically wrong conclusions to which one can be led by assuming θ to be a proper variable is given by R. E. Peierls and J. N. Urbano, *Proc. Phys. Soc.*, 1, 1, 1968.

It should be stressed that the difficulty expressed in (1.4.4) depends on the fact that M has discrete eigenvalues, which in turn depends on the periodicity in θ. In the case of a Cartesian coordinate, x, the conjugate momentum, p_x, has a continuous spectrum, and the matrix representation of the commutator then involves the Dirac δ-function and its derivative, as discussed by Dirac.

Caution may, however, be necessary if one uses a cyclic boundary condition for x, as is often convenient in the electron theory of metals. Then x is no longer a single-valued, continuous variable, and no operator for x exists. One would like to have such an operator to represent the potential of a static electric field in conductivity calculations. The difficulty here has to do with the impossibility of maintaining a static e.m.f. in a circular wire. We could replace the cyclic boundary condition by "rigid-box" condition, but this is no good for conductivity calculations, since no steady current can then flow through the box. In fact, the relations which one often sees derived by the improper use of a linear static potential and a cyclic condition can be justified by more careful reasoning. This is also true of our use, in section 1.3, of the field phase as a variable, since it is easy to replace the uncertainty relation (1.3.3) by one involving the uncertainty in $e^{i\gamma}$.

1.5. THE ADIABATIC EXPANSION

We consider a quantum system whose Hamiltonian contains a parameter, f, which varies slowly with time. This parameter could, for example, represent a slowly varying external electric or magnetic field. The requirement of "slow" variation will turn out to mean that the time variation of f should not cause a substantial variation of the Hamiltonian in a time of the order of the natural periods of the system with constant f.

It is then convenient to introduce the eigenfunctions $u_n(f, x)$ and the eigenvalues $E_n(f)$ of the Hamiltonian $H(f)$, regarding f as a parameter:

$$[H(f) - E_n(f)]u_n(f, x) = 0. \tag{1.5.1}$$

Here x stands for all the independent variables of the problem. We then expand the time-dependent solution of the Schrödinger equation:

$$\psi(x, t) = \sum_n a_n(t) u_n[f(t), x].$$

(1.5.2)

Inserting this in the time-dependent Schrödinger equation, we find

$$\dot{a}_n = -\frac{i}{\hbar} E_n[f(t)] a_n + \sum_m W_{nm}[f(t)] \dot{f} a_m,$$

(1.5.3)

where

$$W_{nm} = \int u_n(f, x) \frac{\partial u_m(f, x)}{\partial f} dx.$$

(1.5.4)

We can, as usual, eliminate the first term on the right-hand side of (1.5.3) by defining

$$a_n = b_n \exp\left(-\frac{i}{\hbar} \int^t E_n[f(t')] dt'\right)$$

(1.5.5)

with some arbitrary lower limit to the integral.
 Then

$$\dot{b}_n = \dot{f} \sum_m W_{nm} \exp\left(-\frac{i}{\hbar} \int^t \{E_m[f(t')] - E_n[f(t')]\} dt'\right) b_m.$$ (1.5.6)

Evidently the coefficients b_n are time-independent if f is constant, and, to the leading order in the rate of change of f, the system remains in the same quantum state in which it started, with only the phase changing by (1.5.5). This is the content of the adiabatic theorem.
 To obtain the next correction, it seems plausible to treat the b_m on the right-hand side of (1.5.6) as constant, since their rate of change is proportional to the small quantity $\dot{f}$, and they appear in the equation multiplied by $\dot{f}$.

Assuming, for definiteness, that the system started at $t = -\infty$ in some initial state i, then in the approximation described:

$$\dot{b}_n = \dot{f} W_{ni} \exp\left(-\frac{i}{\hbar} \int^t \{E_i[f(t')] - E_n[f(t')]\} dt'\right), \quad (1.5.7)$$

which reduces the problem to a quadrature.

In most applications one is particularly interested in the probability that the system will be found in a state n after the field has ceased acting. In that case we require the integral of (1.5.7) over all times from $-\infty$ to $+\infty$. For a smoothly varying field this integral can be extremely small, decreasing exponentially with the time scale of the variation of f.

Consider, by way of illustration, the case of the harmonic oscillator in a time-dependent electric field. Then

$$H = \frac{1}{2m} p^2 + \frac{m}{2} \omega^2 x^2 - m\omega^2 f x$$

$$= \frac{1}{2m} p^2 + \frac{m}{2} \omega^2 (x - f)^2 - \frac{m}{2} \omega^2 f^2. \quad (1.5.8)$$

The ground state and first excited states are

$$u_0 = \left(\frac{\beta}{\pi}\right)^{1/4} \exp\left(-\frac{\beta}{2}(x - f)^2\right),$$

$$u_1 = 2\sqrt{\beta}\left(\frac{\beta}{\pi}\right)^{1/4} (x - f) \exp\left(-\frac{\beta}{2}(x - f)^2\right), \quad \beta = \frac{m\omega}{\hbar}$$

so that

$$\frac{\partial u_0}{\partial f} = \frac{1}{2}\sqrt{\beta}\, u_1.$$

Hence W_{n0} as defined by (1.5.4) vanishes unless $n = 1$. The energies are

$$E_n(f) = (n + \tfrac{1}{2})\hbar\omega - \tfrac{1}{2}m\omega^2 f^2.$$

For a system which is in the ground state at $t = -\infty$, eq. (1.5.7)
now reads:

$$\dot{b}_1 = \dot{f}\tfrac{1}{2}\sqrt{\beta}\,e^{i\omega t} \tag{1.5.9}$$

As an example, choose

$$f = A e^{-\frac{1}{2}\gamma t^2} \tag{1.5.10}$$

The probability of the system being in the first excited state when
the field has ceased acting is $|b_1(\infty)|^2$. Integrating (1.5.9) over all
times we find

$$b_1(\infty) = i\omega A \left(\frac{\pi\beta}{2\gamma}\right)^{1/2} e^{-\frac{1}{2}\omega^2/\gamma}. \tag{1.5.11}$$

The important point is that this is very small if the adiabaticity
parameter γ/ω^2 is small. In fact, the expression decreases faster
than any power of the parameter. Now in going from (1.5.6) to
(1.5.7) we have neglected terms of second order, i.e., containing the
square of γ. Evidently it does not follow that such terms are small
compared with our result, and that they may not be important.

It might be thought that the smallness of (1.5.11) is due to the
choice of a Gaussian for $f(t)$. To show that that is not so consider
the Lorentzian form

$$f = \frac{A}{1 + \gamma t^2} \tag{1.5.12}$$

The time integration can again be done in closed form:

$$b_1(\infty) = -\frac{i\pi\omega}{2}\sqrt{\frac{\beta}{\gamma}}\,A e^{-\omega/\sqrt{\gamma}}. \tag{1.5.13}$$

which is again exponentially small.

It is difficult to obtain a more reliable estimate of the small transition probability in the nearly adiabatic case. Some numerical studies in the 1970 Oxford DPhil thesis by J. B. Scheffler, while not covering the subject exhaustively, suggest that the exact answer, like the "approximate" answer obtained from (1.5.7), is exponentially small, but not necessarily approximated by (1.5.7).

It is instructive to compare this problem with one of similar structure, for which an exact answer is known in a special case. That is the problem of a particle in one dimension passing a smooth potential barrier or well, in which the potential varies slowly compared to the wavelength.

We write the Schrödinger equation for this problem in the form

$$\frac{d^2\psi}{dx^2} + [k^2 - u(x)]\psi = 0 \qquad (1.5.14)$$

where k is the wave number, and $u(x)$ is the potential, apart from a scale factor $2m/\hbar^2$. We write

$$\psi(x) = a_1(x)e^{i\phi(x)} + a_2(x)e^{-i\phi(x)}, \qquad (1.5.15)$$

where

$$\phi(x) = \int^x \sqrt{[k^2 - u(x')]}\, dx' \qquad (1.5.16)$$

is the exponent occurring in the W.K.B. approximation.

If the coefficients a_1 and a_2 in (1.5.15) were constant, we would have just the leading order of the W.K.B. approximation, and, provided ϕ is always real, i.e., the potential of the barrier is less than the energy of the particle, there would be no reflection in that approximation. To improve the approximation we allow a_1 and a_2 to vary. The definition (1.5.15) does not determine them uniquely, and we can conveniently add the requirement that

$$\psi' = i\phi'(a_1 e^{i\phi} - a_2 e^{-i\phi}). \qquad (1.5.17)$$

From the Schrödinger equation and our definitions we then find

$$a_1{}' = -\frac{1}{2}a_1\frac{\phi''}{\phi'} + \frac{1}{2}a_2\frac{\phi''}{\phi'}e^{-2i\phi},$$

$$a_2{}' = -\frac{1}{2}a_2\frac{\phi''}{\phi'} + \frac{1}{2}a_1\frac{\phi''}{\phi'}e^{2i\phi}. \tag{1.5.18}$$

We can eliminate the first term on the right-hand side of each of these equations by setting

$$a_{1,2} = b_{1,2}(\phi')^{-1/2} = b_{1,2}(k^2 - u)^{-1/4}. \tag{1.5.19}$$

Constant b_1 and b_2 would correspond to the second-order W.K.B. approximation, with still no reflection. Inserting in (1.5.18):

$$b_1{}' = \frac{1}{4}\frac{u'}{k^2 - u}e^{-2i\phi}b_2,$$

$$b_2{}' = \frac{1}{4}\frac{u'}{k^2 - u}e^{2i\phi}b_1. \tag{1.5.20}$$

The structure of these equations is very similar to that of (1.5.6) for a case in which only two levels contribute, except that in that case, if the states are non-degenerate, we may take the eigenfunctions as real and then from (1.5.4) $W_{12} = -W_{21}$. This leads to a pair of equations with opposite signs. However, it is not likely that this will make a major difference in the behavior of the solutions.

For the reflection problem of equation (1.5.14), an exact solution is given in Landau and Lifshitz, *Quantum Mechanics* (Problem 4 in §23) with the potential

$$u(x) = -\frac{u_0}{\cosh^2 \alpha x} \tag{1.5.21}$$

They find the transmission coefficient

$$T = \frac{\sinh^2 \pi k/\alpha}{\sinh^2 \pi k/\alpha + \cos^2 \left[\frac{\pi}{2} \left(1 - \frac{4u_0}{\alpha^2} \right)^{1/2} \right]}. \qquad (1.5.22)$$

We are interested in the adiabatic limit, in which k/α is large. Then, to the leading order, we find for the reflection coefficient, R,

$$R = 1 - T \sim 4 \cos^2 \left[\frac{\pi}{2} \left(1 - \frac{4u_0}{\alpha^2} \right)^{1/2} \right] e^{-2\pi k/\alpha}. \qquad (1.5.23)$$

If, in addition to the assumption of small α/k, we assume $4u_0/\alpha^2$ to be a small quantity, this simplifies to

$$R = \frac{4\pi^2 u_0{}^2}{\alpha^4} e^{-2\pi k/\alpha}. \qquad (1.5.24)$$

If, on the other hand, we solve (1.5.20) in the same spirit as (1.5.7), we take the forward amplitude b_1 to be constant, equal to unity, and determine $b_1(-\infty)$ as the amplitude of the reflected wave. If we again assume u_0 small, so that the denominator of (1.5.20) becomes k^2, and ϕ in the exponent becomes kx, $R = |b_2(-\infty)|^2$ becomes identical with (1.5.24).

Thus, in this particular problem, the first term of the adiabatic expansion gives the correct answer if the potential is slowly varying *and* weak. The evaluation of the adiabatic approximation is more troublesome if the potential is not weak, and I do not know whether the answer would still be right. (This does not seem likely, because in evaluating (1.5.20) we neglected terms of the relative order u_0/k^2, whereas the corrections to (1.5.24) are of the order u_0/α^2.) It is also not known whether the agreement is a feature of the particular potential used, or is more general.

1.6. INTERPRETATION OF QUANTUM MECHANICS

The two great revolutions in physics in the twentieth century, relativity and quantum mechanics, required drastic revisions in our basic concepts of space, time, and causality. The ideas of relativity, which affected particularly the concept of time, originally met with considerable opposition from philosophers. One possible reason for this was the fact that the word "relativity", in the sense that "all is relative", had been used in philosophy to describe an attitude very different from that of the theory of relativity. Indeed the essence of the latter is to formulate the laws of physics in a form in which they have absolute validity, independently of the state of motion of the observer. Very probably some of the early opposition was not really based on the content of the new theory, which was difficult to grasp, but on the name.

As a better appreciation of the nature and content of the theory developed, it found more general acceptance, and today one does not find many philosophers, or physicists, who see any conceptual objection to the ideas of relativity.

In the case of quantum mechanics, the objections from philosophers were at first far less vocal. In the late nineteen-twenties, and early thirties, the conceptual revolution characterized by the duality between waves and particles, and by the uncertainty principle, was a subject of intense discussion among physicists, who had to adjust themselves to the new ideas, refine them, and satisfy themselves about their consistency. Born, Heisenberg, and Niels Bohr led the formulation of consistent principles. Others, including Einstein, who had contributed much to the development of quantum theory and even to its interpretation, were reluctant to abandon the classical and intuitively convincing ideas of causality and determinism. Einstein made many attempts to find counter-examples to the uncertainty principle; invariably Niels Bohr found the flaw in the argument; and these debates helped to deepen our understanding of the new ideas and strengthen our confidence in them.

Although Einstein remained unconvinced that quantum mechanics, and the uncertainty principle, which is an essential part of it, was inescapable, he respected the theory and did not underestimate the magnitude of the task of modifying it or replacing it by another system which would leave its many successes intact.

Schrödinger also was unhappy about the loss of the simple and attractive picture in which the wave function of an electron in the hydrogen atom was a physical quantity like the vibration of a string, or the field in a waveguide. He had to admit, however, that this simple picture could not account for phenomena including more than one electron, and that it contradicted the observed facts in many other ways. Although the accepted interpretation of quantum mechanics made him uncomfortable, he did not claim that he could suggest another. De Broglie at one time came to the same conclusion; in a book published in 1929, he convinced himself and the reader that the alternative, deterministic forms of the theory, which at first sight looked attractive, were not acceptable. In later years he returned to the search for alternative forms of the theory.

The general community of physicists had to go through the same arguments, and to face the same conceptual difficulties, which result from the fact that our intuition is formed as a result of our everyday experience on a scale on which neither the relativistic refinements nor the quantum effects are noticeable. Most of them learned to accept the new ideas, and to understand that they are an essential part of the logic of quantum mechanics.

One specific question of interest in connection with the basis of the theory is whether there could be "hidden variables." In the observation of a physical variable which is capable of two values, to take a simple case, quantum mechanics cannot, in general, predict the outcome, but only give the probability of finding either result. In classical physics we also meet situations in which we can make only probability statements about the outcome of an experiment, notably in statistical mechanics, but that happens because we are in practice unable, or unwilling, to measure all the variables which would determine the outcome. The question is then: Can we imagine, in quantum mechanics, the existence of variables whose values are unknown to us, and by the uncertainty principle must remain unknown, but whose knowledge *would* determine the result of the experiment?

Leaving aside the question of what we would gain from the

knowledge of the existence of such hidden variables if we could never know their values, the problem seemed settled when von Neumann gave a proof that any theory whose observable predictions would always agree with those of quantum mechanics could not contain hidden variables.

In recent years the debate on these ideas has reopened, and there are some who question what they call "the Copenhagen interpretation of quantum mechanics"—as if there existed more than one possible interpretation of the theory.

I want here to comment on two aspects of this debate, which contain elements of surprise. The first is that, as a result of such discussions, J. S. Bell had occasion to re-examine the proof by von Neumann to which I have referred. He discovered that this proof relied on an assumption for which there seemed physically no compelling reason, and that, without that assumption, the theorem stated by von Neumann was not valid. Bell was, however, able to show that, to give a correct account of certain important types of measurement, hidden variables must violate a condition of locality, which does seem an essential physical requirement. The position has thereby been restored to what it seemed to be after von Neumann's proof. The element of surprise in this situation is the fact that such a vital point in an argument cited so frequently could have remained unnoticed for so long.

The work of Bell is contained in two papers: the first part in *Rev. Mod. Phys.*, 38, 447, 1966, the second in *Physics*, 1, 195, 1964. (Note that the second part was published about two years earlier than the first.)

In discussing Bell's argument it must be understood that, like von Neumann's, it relates only to the possibility of introducing hidden variables while leaving all observable predictions of quantum mechanics unchanged. We are not here discussing the alternative possibility of modifying the theory so as to make the introduction of hidden variables possible. To be acceptable, such a modified system of mechanics would have to reproduce all those predictions of quantum mechanics which are well confirmed by observation, and in the domain of non-relativistic mechanics this covers a wide range of phenomena. It therefore seems extremely difficult to visualize such a modified theory, but it does not seem possible to prove that such an unspecified modification cannot exist. In the field of particle physics we have as yet no guarantee that the existing principles of quantum mechanics are adequate for a complete description, although no definite contradiction has been found. If new concepts are needed here, as is quite possible, one might guess that they would take us yet further away from classical concepts and from determinism, rather than back toward them. But all this must remain a field for speculation and for the future.

Let us then return to the more concrete possibility of hidden variables supplementing a valid quantum mechanics. Consider two operators A and B, belonging to two observables. Then $A + B$ is also an observable with an associated operator. In classical physics, an observation of $A + B$ could be carried out by observing A and B separately and adding the results. In quantum physics this is barred by the uncertainty principle, unless A and B happen to commute. A measurement of $A + B$ is then a separate experiment, which does not involve measuring either A or B. Nevertheless, according to the basic postulates of quantum mechanics there is still a linear relation between the expectation values:

$$\langle A + B \rangle = \langle A \rangle + \langle B \rangle. \tag{1.6.1}$$

If there exist hidden variables, there would be some quantity η (which could comprise several variables) whose knowledge would determine the results of all observations on the system, including A, B, and $A + B$. Von Neumann takes it for granted that the values predicted for given η will still satisfy the same relationship:

$$(A + B)(\eta) = A(\eta) + B(\eta) \tag{1.6.2}$$

But there is no reason why this is necessary. All that follows from (1.6.1) is that the left and right-hand sides of (1.6.2) have the same average if we average over all η with suitable weights.

Once the general validity of (1.6.2) is assumed, it is very easy to arrive at a contradiction. Consider, in the case of an object of spin $\frac{1}{2}$, the spin component in some direction in the x,y plane,

$$s' = s_x \cos \theta + s_y \sin \theta. \tag{1.6.3}$$

Quantum mechanics predicts that s_x, s_y, and s' are each capable only of taking the values $\pm\frac{1}{2}$. In other words, we have to find two functions $s_x(\eta)$, $s_y(\eta)$ such that each is $+\frac{1}{2}$ for some values of η and $-\frac{1}{2}$ for others, so that the combination (1.6.3) is also always $\pm\frac{1}{2}$. This is manifestly impossible. So, on the basis of this postulate, von Neumann's theorem follows immediately.

However, without this postulate, there is no contradiction. To show this, Bell constructed an example of functions s_x, s_y, s' of a variable η, which took only values of $\pm\frac{1}{2}$ and for which (1.6.3) is satisfied on the average. There is no claim that this rather contrived construction has any physical basis, but it serves to prove that von Neumann's postulate is essential for his result.

In his second paper, Bell was able to show that, for more complicated situations than observations on a single spin, the introduction of hidden variables would require a property which he calls "non-locality." This arises in the following way: Consider the experiment discussed in the paper by Einstein, Podolsky, and Rosen (*Phys. Rev.*, 47, 777, 1935). Two objects of spin $\frac{1}{2}$ are known to be in a state of resultant spin 0. If we then measure the x component of the spin of particle 1, and find it to be $+\frac{1}{2}$, we can predict with certainty that the x component of the spin of particle 2 must be found to be $-\frac{1}{2}$. The same applies if we choose to measure the spin of particle 1 in any other direction. We note that this strong correlation remains valid at later times, provided only that the particles have not been exposed to any magnetic fields or other spin-dependent interactions. During this time the particles may have traveled long distances, and may be well separated in space.

Bell shows that the hidden variables cannot reproduce all the strong predictions of quantum mechanics for this situation unless, for given value of the hidden variable, the spin of the second particle still depends on which spin component of the first we have chosen to observe. Since the particles may be so far from each other that no signal can get from one to the other in the time between the observations on particle 1 and the related observation on particle 2, Bell rightly concludes that such a non-local influence is physically unreasonable.

We therefore restore the conviction, apparently justified by von Neumann's proof, that hidden variables cannot be introduced without abandoning some of the results of quantum mechanics, if we replace von Neumann's postulate of additivity by Bell's more physical postulate of locality.

The other comment I want to make relates to the step which is often referred to as the "contraction of the wave packet." Consider a quantum system described initially by a wave function ψ. If we carry out an observation on the system, we cannot in general predict the result with certainty, but give only probabilities for the various possible outcomes. This means that ψ can be thought of as a superposition, with suitable coefficients, of the eigenfunctions ϕ_n belonging to the eigenvalues a_n of the observable A which is to be measured. As soon as the observation has been carried out, yielding, say, the value a_1 for A, all parts of the wave function other than ϕ_1 have to be discarded. If the observation is one of position, ϕ_1 will be much more closely localized in space than ψ (to a mathematical point in the limit of an accurate observation of position) and this is the origin of the term "contraction"; however, ϕ_1 is of course much more diffuse in momentum space than ψ.

It is sometimes thought that this contraction would come out as a consequence of the time-dependent wave equation, if the interaction between the system under observation and the measuring apparatus were properly taken into account. If this were true, it would give rise to grave conceptual difficulties. For example, in the Einstein-Podolsky-Rosen experiment discussed above, the wave equation should cause a change in the wave function affecting predictions about particle 2, as a result of the interaction of an instru-

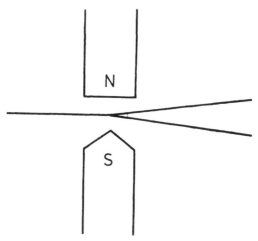

Figure 1.3   Stern-Gerlach experiment.

ment with particle 1. If the particles are by this time distant in space, this would take some time, since the wave equation, which is consistent with relativity, cannot allow the propagation of any influence with a speed exceeding that of light.

However, this expectation is quite wrong, as we can see by looking a little more closely at a typical process of measurement. For this we choose a Stern-Gerlach experiment as example. Figure 1.3 illustrates this schematically. A beam of atoms of spin $\frac{1}{2}$ passes through a region with an inhomogeneous magnetic field. On the right we then see two beams, one containing the atoms with $s_z = +\frac{1}{2}$, the other those with $-\frac{1}{2}$.

It is very easy to visualize the solution of the Schrödinger equation for this problem. The incident wave has two components, in general of comparable amplitude, for the two values of s_z. This is sketched in Fig. 1.4a. After passing through the magnet gap, the two components are peaked around different values of the transverse coordinate, y, as in Figure 1.4b, representing the fact that the deflection in the y direction depends on the spin.

We see that no contraction has taken place in this "measurement". Both spin directions are still equally represented, and the only new

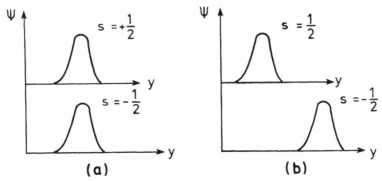

Figure 1.4   Spin and space dependence of wave function: (a) before, (b) after "measurement."

feature is a correlation between the position of the atom and its spin component. In this sense all a measuring instrument ever does is to generate correlations.

The description of the Stern-Gerlach experiment is, however, not yet complete, because, having established the correlation, it remains to find whether the atom is in the upper or the lower beam. We can do this, for example, by placing counters in the two beams, sensitized to the atoms in question. We can, if we wish, include the operation of the counters in our quantum treatment, introducing new degrees of freedom to describe the state of the counters. We need not carry out the calculation, because it is clear what will happen. The new wave function, defined in the space containing the spin and position of the atom, and the two counters, will in its final form consist of two parts: one for spin up, atom in upper beam, upper counter activated, lower counter not activated; and the other with spin down, atom in lower beam, and only the lower counter activated. We now have set up a threefold correlation.

We can pursue this through the galvanometer or other visible indicator for the state of the counters, the light reflected from the indicator as we look at it, the action of this light on our retina, and so on. All this would do is to generate more and more correlations, without discarding either of the original options. The reason we do not normally discuss this chain of events is that beyond a certain point, in our case from the operation of the counters on, quantum effects, and the operation of the uncertainty principle, are negligible, so that we may assume the correlations in the later links of the chain as perfect.

When do we contract the wave function? Not until we have become aware through our visual, or other sensory observation, that one or the other of the two alternative possibilities is realized. Once we *know* that the atom has activated, say, the upper counter, we conclude that its spin was up, and we discard, from now on, that part of the wave function which contains an atom with $s_z = -\frac{1}{2}$.

This situation sounds extremely unsatisfactory if one tries to take the view that the wave function, or in the more general case the state function, should be a physical object, like an observable. But this is a mistaken view. Any careful assessment of the process of measurement in quantum mechanics, as sketched above, makes it clear that the significance of the state function is to represent our state of knowledge of the system. Once this is accepted, it is not surprising that new information can change the state of our knowledge. It is also clearly no difficulty that a measurement performed in one region of space can give us information about an object which is far away, without this implying the transmission of influence instantaneously.

One question we may ask at this point is: Whose knowledge does the wave function represent? Do we have to write a separate state function for every observer who participates in an experiment, and perhaps for every reader of the resulting paper? The answer is that in most practical experiments, as in the example of Stern-Gerlach, the later links in the chain leading from the quantum system to the reading of an instrument, are entirely classical. Many observers may therefore look at the same dial without affecting the state of affairs; they can pass information about what they have seen, and verify each other's statements. In these circumstances we are then dealing with common knowledge, and no distinction need be made.

There are exceptions to this. We can imagine an observation being conducted entirely within the quantum domain. It is known that the sensitivity of the human eye can come very close to detecting a single light quantum, and we can certainly imagine a case in which an observer sees a single photon. Then only one person acquires the knowledge in the first place, though he, too, can share the knowledge gained with others. No new difficulty of principle arises.

Another objection which is sometimes raised is the following: Suppose we include in our quantum description the observer and his brain. In practice, the resulting Schrödinger equation would be prohibitively complicated even to formulate, let alone to solve, but it should be possible in principle. What is then the significance of the state function occurring in this equation; whose knowledge does it represent, and when do we decide to contract it?

This question seems to pose an unsurmountable difficulty. But it is based on the assumption that living beings, such as our observer, can be described by the existing laws of physics, in other words, that biology is ultimately a branch of physics in the sense in which chemistry is now known to be, in principle, a branch of physics. We are confident today that, if we could solve the Schrödinger equation for all the electrons in a large molecule, it would give us all the knowledge that chemists are able to discover about it. (This does not make the chemist's job redundant, because the solution of such equations is far beyond the range of the most powerful computer we can imagine.)

Many people take it for granted that the same must be true of the science of life. The difficulty about how to formulate the acquisition of information, which we have met, is a strong reason for doubting this assumption.

This is not far from the question of how one would incorporate the concept of consciousness into a description of living beings in terms of the physical functioning of their brain cells. Consciousness is admittedly hard to define objectively, but each of us has a clear intuitive understanding of what he means by being conscious. One can compare the human brain with a very sophisticated computer, and indeed a computer can perform many of the functions of the brain, but it does not seem easy to imagine a computer being conscious. This problem is far from physics, but it does connect with the argument to which we have been led, because knowing that a measurement has disclosed a certain event is the same thing as becoming conscious of the fact, and this is precisely what makes us contract the state function.

In claiming that biology is not likely to be a branch of the present physics, I do not wish to imply that life can in some mysterious way evade the laws of physics. I believe that the situation is comparable to the problem of electricity and magnetism as it appeared before, and even during, the time of Maxwell. Physics then was mechanics, and to explain a phenomenon meant to find its mechanism. Even Maxwell tried hard to back his field equations by mechanical models. He understood only later that electric and magnetic fields were important basic concepts in physics, not contained in the concepts of mechanics, but of course not in contradiction to the laws of mechanics—and that physics had to be enriched by adding them. It is at least possible, and to me probable, that similarly new concepts have to be added to our present physical ones before an adequate description of life is possible. Whether the thus enlarged discipline should still be called physics is a semantic question.

1.7. γ-RAY MICROSCOPE

We turn to a very much simpler topic, still connected with the uncertainty principle of quantum mechanics. It starts with an anecdote.

When Heisenberg, then a student in Munich, submitted his PhD dissertation, he was already known as a young man of outstanding ability. But he had aroused the displeasure of W. Wien, the professor of experimental physics, by not taking the laboratory classes seriously enough. It was then part of the requirements for the PhD to take a quite searching oral examination in the relevant subjects. When Heisenberg submitted himself to questioning by Wien, the first question related to the resolving power of a microscope, and the candidate did not know the answer. The next question was about the resolving power of a telescope, and he still did not know. There were more questions about optics and no answers, and the professor decided the candidate should fail. However, a joint mark had to be returned for both experimental and theoretical physics, and after difficult negotiations between A. Sommerfeld, the theoretical physicist, and Wien, Heisenberg passed in physics with the lowest pass mark.

Four years later he wrote his famous paper about the uncertainty principle. As one of the main illustrations of this he discusses a hypothetical microscope using γ-rays so as to improve the resolving power and have, in principle, an observation of the position of an electron with an accuracy on the atomic scale. He then shows that the photon, whose scattering by the electron is used to detect it, will communicate momentum to the electron. The amount of momentum is greater the shorter the wavelength of the photon, and a short wavelength is required for good resolving power. The momentum transfer can result in an uncertainty in the momentum of the electron, even if its momentum before the event was accurately known.

Quantitatively, Heisenberg just stated that the resolving power was of the order of the wavelength, so that the error in the determination of the position was at best of the order of λ, which is also h/p, where p is the momentum of the photon. If the photon is deflected, the momentum change will be of order p, and so will be the uncertainty in the final momentum of the electron, Δp. Hence the errors in x and p are related by

$$\Delta x \, \Delta p \gtrsim h. \tag{1.7.1}$$

However, as was pointed out by Niels Bohr, the recoil momentum is not necessarily completely unknown. One knows the direction of the incident light, hence the initial momentum of the photon, and one also knows that it entered the lens of the microscope, so one knows something of its final direction. Particularly if the microscope has a small aperture, one might therefore achieve a momentum uncertainty much less than p.

The way out, also noted by Bohr, is that the formula for the resolving power is actually

$$\Delta x = \frac{\lambda}{\sin \frac{1}{2}\theta}, \qquad (1.7.2)$$

where θ is the angular aperture. On the other hand, $p \sin \frac{1}{2}\theta$ is just the uncertainty in the momentum arising from the fact that we know its final direction only within the aperture. Thus (1.7.1) is justified.

Heisenberg added a note to the proof of his paper, giving this extra argument to establish the relation (1.7.1). The surprise in this case must have been Heisenberg's, but the story provides an amusing illustration of the fact that even for a great man a sound knowledge of old-fashioned physics can be essential.

1.8.   ALARM-CLOCK PARADOX

Our next topic also relates to the implications of the uncertainty principle. As I mentioned already, Einstein was, at one time, trying to show the inconsistency of the uncertainty principle by inventing experiments which would result in a determination of physical variables with greater accuracy than the uncertainty principle would permit. One of his suggestions concerned the relation between energy and time:

$$\Delta E \, \Delta t \gtrsim h. \qquad (1.8.1)$$

The significance of this relation is not as simple as, for example, that between position and momentum. This is because time is not an observable. A measurement of time in itself does not convey any information about a physical system, and a statement about any other physical quantity usually implies that we are talking about its value at some particular time. In the case of a conserved quantity, such as the energy of an isolated system, the result then also gives the energy at any time. Landau was fond of making this point by saying: "There is evidently no such limitation—I can measure the energy, and look at my watch; then I know both energy and time!"

There are, however, a number of meaningful questions, to which the inequality (1.8.1) provides the answer. One of these relates to the transfer of energy from one part of a physical system to another. If ΔE is the error with which we know the transferred amount of energy, and Δt the error in our knowledge of the time when this energy transfer took place, then the inequality is valid.

To disprove this, Einstein proposed the following arrangement: A box with perfectly reflecting walls contains some radiation, which remains there, with constant total intensity, as long as the box remains closed. The box carries a clockwork device which is set to open a window in the wall of the box at a pre-arranged time for a short interval Δt. This would seem to fix, with an error of Δt, the time at which any radiation can have escaped from the box. The amount of radiation emitted can be determined by finding the total energy content of the box both before and after the operation of the shutter, by weighing the box carefully and using the relation $E = mc^2$ between the mass of the box and its energy content. There would seem to be no connection between Δt and the accuracy of the energy determination.

Einstein put this idea to Niels Bohr, who, after considerable reflection, came up with the following solution. Weighing an object means observing the force exerted on it by gravity, and the only way of measuring a force is by the application of Newton's law. In other words, we have to measure the rate of change of the total momentum in the field of gravity. (If this sounds an unfamiliar description of the process of weighing, remember that the usual form is to wait for the scales to come into equilibrium, by allowing friction to dissipate the momentum gained. However, in checking that the scales have now come into equilibrium we must still check that the rate of change of momentum is zero.)

Let the box accelerate under gravity for a time t_1, and assume we measure the momentum gained with an error Δp. Then we know the force of gravity on the box with an error of $\Delta p/t_1$, the mass with an error $\Delta p/t_1 g$, where g is the gravitational acceleration, and hence the energy with an error of $\Delta p c^2/g t_1$.

The knowledge of the momentum of the box to within Δp implies, by the usual form of the uncertainty principle, a lack of knowledge of its position to better than Δx, which is connected with Δp in the usual way. Now comes the crucial point: By the general theory of relativity, or simply by the principle of equivalence, the time measured on a clock is affected by a correction proportional to the gravitational potential at the location of the clock. During the weighing, the position of the box, and hence of the clock in it, is unknown by Δx, and therefore the gravitational potential by $g\Delta x$. This causes an unknown error in the rate of the clock of $g\Delta x/c^2$, and after the time t_1 the reading of the clock is in error, relative to the time of an outside observer, by

$$\Delta t = t_1 g \,\Delta x/c^2 . \tag{1.8.2}$$

Combining this with our result for ΔE, and using the relation between Δp and Δx, we recover (1.8.1), since the error in the clock reading is also an error in our knowledge of the time when the shutter, was, in fact, opened. Note that we cannot determine the error in the clock reading by checking the clock afterwards, because the experiment requires us to measure the energy of the box also after the emission of the radiation, and therefore we need a second weighing, which causes another error in the clock reading of the same order as the first. We cannot therefore find out how much the clock was in error between the two weighings, at the time it operated the shutter.

One could ask whether the experiment could be improved by not mounting the clock on the box, but keeping it in a fixed place and having it operate the shutter by remote control. In that case errors are introduced not in the reading of the clock, but in unpredictable delays in the transmission of the signal from the clock to the shutter, which sits in the gravitational field at an undetermined position.

Our surprise, no doubt shared by Einstein, comes here from the way in which the gravitational redshift of general relativity comes in to reconcile what would otherwise be a paradox in the application of the identity $E = mc^2$ of special relativity to the uncertainty relation of quantum mechanics.

Let us note that other implication of the relation (1.8.1) are not without difficulty. One implication that is usually accepted is that a measuring apparatus which interacts with the system to be studied only for a time Δt, cannot determine the energy of the system to better than ΔE, the two errors again related by the same inequality. To this rule, Aharonov and Bohm (*Phys. Rev.*, 122, 1649, 1961) have given a counter-example. In this, admittedly very academic, example, the interaction is given by a term in the Hamiltonian proportional to the momentum of the particle representing the "system" and the coordinate of the "instrument" (or vice versa) with a time-dependent factor which vanishes outside a time interval of length Δt.

The likely solution to this paradox is that this interaction Hamiltonian for some reason cannot be permitted in quantum mechanics, although nobody has seen what general principle it violates. It is perhaps significant that a very similar interaction term, namely,

$$f(t)(xp_y - yp_x), \tag{1.8.3}$$

can occur in the interaction of a charged particle with a time-dependent magnetic field. Each of the two terms in this expression is of the form used by Aharonov and Bohm, if one is willing to regard one coordinate of the particle as the "system" and the other as the "instrument," but if both terms are present, as in the real physical case, the result no longer violates the inequality (1.8.1). There is still room for another surprise here.

2. QUANTUM THEORY OF ATOMS

2.1. SCATTERING OF γ-RAYS

If an atom is exposed to radiation of a frequency much higher than the resonance frequency of even its K electrons, the dominant processes are Compton effect and photo-ionization. However, for some purposes one is interested in the coherent scattering, in which the atom remains in its ground state, although the cross section for this is rather small.

If we use Dirac's equation for describing the atomic electron, the scattering of radiation is a two-stage process in which first the incident photon is absorbed, and then the secondary photon emitted, or vice versa.

The contribution to the scattering amplitude of the first process is proportional to

$$\sum_n \frac{\langle \psi_0 | e_\mu' \gamma_\mu \, e^{-i\mathbf{k}\cdot\mathbf{r}} \, | \psi_n \rangle \langle \psi_n | e_\nu \gamma_\nu \, e^{i\mathbf{k}\cdot\mathbf{r}} | \psi_0 \rangle}{E_0 + \hbar c\omega - E_n} \tag{2.1.1}$$

where e_μ, e_ν' are the polarization vectors and $\mathbf{k}$, $\mathbf{k}'$ the wave vectors of the incident and scattered photon, respectively, γ_μ the Dirac matrices, ψ_0 the atomic wave function for a K electron, and ψ_n the wave function of any excited state. The summation over n includes integration over the continuous spectrum. So far this result is exact, except for radiative corrections of the relative order of magnitude $\alpha = e^2/\hbar c$, if we include also the similar term in which the emission precedes the absorption.

For high k, the matrix elements in (2.1.1) favor highly excited states of large positive energy, and for these it seems a plausible approximation to neglect the potential energy, and to replace the scattering states by plane waves. This approximation should be particularly adequate for our purpose when the photon energy is high and the atom light, so that the Coulomb potential, which we are neglecting, is weak. However, one finds here what is almost the classical case of a theoretical surprise, because it turns out that the result for the scattering amplitude obtained this way is wrong, and gets worse the higher the photon energy and the lighter the atom.

Indeed the evaluation of the integration in (2.1.1) and in the other term is straightforward if one neglects also the binding energy of the initial state compared to the photon energy. The result is well known, and is quoted, for example, in the paper by Brown and Woodward, *Proc. Phys. Soc.*, A65, 977, 1953, in which this surprise was first noticed.

To check the approximation, they calculate also the next term obtained, by adding, to the plane waves for the intermediate states, the first-order Born correction for the scattering of a free electron by the Coulomb field. They find then for the ratio of the first-order term to the approximation which uses only plane waves,

$$\frac{A_1}{A_0} = \frac{\Delta p}{\pi Z \alpha mc} \frac{1}{1 + \dfrac{8mc}{Z\alpha\,\Delta p}}. \qquad (2.1.2)$$

Here $\Delta p = \hbar|\mathbf{k}' - \mathbf{k}|$ is the momentum transfer. In the circumstances we have specified, Δp is much larger than mc, and $Z\alpha$ is a small quantity. Thus the ratio is large, just in the case in which we expected A_0 to be a good approximation. If Δp is very large, and the atom not too light, the last factor is near unity, and the ratio of the correction to the "leading term" is proportional to Δp; for a very light atom and not too large Δp, the second term in the second denominator dominates, and the ratio varies as $(\Delta p)^2$.

As always, it is easy to see the origin of this surprising result, and to understand how it could have been foreseen. The reason is that the "leading term" becomes extremely small. Indeed, if we take for ψ_n a plane wave of wave vector $\mathbf{K}$, the matrix elements occurring in (2.1.1) involve the Fourier transform of the atomic wave function ψ_0 taken for a wave vector $\mathbf{K} - \mathbf{k}'$ and $\mathbf{k} - \mathbf{K}$, respectively. If we choose **K** so as to make one of these arguments small, the other is of order $\Delta p/\hbar$. As a result, the contribution contains the magnitude of the Fourier transform of the atomic wave function at argument $\Delta p/\hbar$. Now the Fourier transform of a non-relativistic hydrogen eigenfunction decreases as the inverse fourth power of the argument; if Δp is large enough we have to use the relativistic form for this tail of the eigenfunction in momentum space, and it turns out that this decreases approximately as the inverse third power of the wave vector.

In the first-order Born correction, on the other hand, we include an additional scattering by the Coulomb potential, and the greatest contribution comes from the term in which this scattering occurs with a momentum transfer of about $\Delta \mathbf{p}$, because then we are left with the Fourier transform of ψ_0 for small argument. This scattering process brings in as a factor the Fourier transform of the Coulomb potential, which is proportional to the inverse square of the wave vector, i.e., to $(\Delta p)^{-2}$. Comparing this with the momentum dependence of A_0 which involves $(\Delta p)^{-4}$, or $(\Delta p)^{-3}$, we understand the behavior of the ratio in (2.1.2). In this discussion I have paid attention only to the dependence on the momentum transfer, for simplicity, but it is equally easy to keep track of the powers of $Z\alpha$ and understand the origin of the Z dependence of the ratio.

The conclusion can be put in more physical terms, as was done in the Brown-Woodward paper, by noting that the momentum transfer Δp must ultimately come from the atomic nucleus. In the approximate expression A_0, the effect of the nucleus is felt only in the ground-state wave function ψ_0, and we therefore rely on the momentum fluctuations contained in that function because of the interation with the Coulomb field. In the correction term we allow the electron to exchange momentum with the nucleus while it is in the excited state, and the comparison shows that it is easier for the electron to pick up momentum when it is already in a state of high energy.

The two terms discussed here are the first two terms in a perturbation series, in which the expansion parameter is the ratio of the atomic potential to the kinetic energy of the intermediate states. We have not looked at the higher terms in this series, but there is little doubt that its convergence will be good when Δp is large and $Z\alpha$ small. The reason we could not neglect the first-order term was not poor convergence, but an extremely small value of the zero-order term. It often happens that the first term in a series expansion vanishes (for example, from symmetry) and then we have no hesitation in going to the next term. In the present case we were in danger of being caught out, because the leading term is non-zero, but smaller than the next one.

While on the subject of coherent scattering of γ-rays, it may be worth recalling that the reason for the interest in the subject in the nineteen-fifties was that there were then experiments in progress to look for Delbrück scattering, which is the scattering of radiation by the Coulomb field of a nucleus, connected with virtual pair creation. For such experiments the scattering by the atomic electrons, which we are discussing, is part of the background and must be allowed for. For light atoms the approach sketched above would be adequate, but for the experiments the heaviest elements were of interest, since they give the strongest Delbrück scattering. In that case the Born series does not converge well for scattering states, even at very high energy. One therefore must evaluate expressions of the form (2.1.1) without approximation.

At first sight this looks a troublesome operation, because the obvious approach would be to obtain the hydrogen wave functions for all discrete and continuous states, compute the matrix elements, and then do the summation/integration. There is a much simpler way, proposed for this purpose by Brown, Peierls, and Woodward, *Proc. Roy. Soc.*, A227, 51, 1954. The device was probably known before, and certainly has been rediscovered many times since.

It consists in writing (2.1.1) in the form

$$\langle \psi_0 | e_\mu' \gamma_\mu \, e^{-i\mathbf{k}\cdot\mathbf{r}} \, | F(\mathbf{r}) \rangle, \qquad (2.1.3)$$

where $F(\mathbf{r})$ comprises all the rest of (2.1.1) and includes the summation over n. It is then easy to show that F satisfies the inhomogeneous wave equation

$$(E_0 - H + c\hbar k)F(\mathbf{r}) = e_\nu \gamma_\nu \, e^{i\mathbf{k}\cdot\mathbf{r}} \, \psi_0(\mathbf{r}), \qquad (2.1.4)$$

where H is the Dirac Hamiltonian.

We have no cause to be surprised that the infinite sum over states representing second-order perturbation theory with respect to the electromagnetic interaction can be reduced to a single inhomogeneous differential equation, because if we remember how to derive the general expressions for perturbation theory in elementary quantum mechanics, we know that this is by way of an inhomogeneous equation of the form (2.1.4). It is usual to expand F in a series of eigenfunctions, for convenience, and this leads to (2.1.1). In this case the raw form (2.1.4) is the more convenient.

It is still a three-dimensional equation, but it can be expanded in a series of angular momentum eigenstates, which converges well if the wavelength $2\pi/k$ is not too small compared with the range of the atomic wave function ψ_0. We are then left with a series of radial equations, which are conveniently solved on a computer.

2.2. LIMITS OF THE HEITLER-LONDON APPROXIMATION

The earliest, and very informative, approach to the quantum theory of molecules was the classical work of Heitler and London on the molecule H_2. Let us recall briefly the idea of the method.

We are concerned with the Hamiltonian for the motion of two electrons in the field of two nuclei, assumed fixed at distance R:

$$H = T_1 + T_2 + U_1 + U_2 + V_1 + V_2 + W_{12} + \frac{e^2}{R}. \qquad (2.2.1)$$

Here the subscript 1 or 2 distinguishes the two electrons, T stands for the kinetic energy, U for the attractive potential of one nucleus, V for that of the other, and W is the interaction between the two electrons.

If the two nuclei are not too close together, the motion of each electron near one nucleus should resemble the ground state of the hydrogen atom. We might therefore try as an approximate solution of the Schrödinger equation the function $u(1)v(2)$, where u and v are the appropriate eigenfunctions of the hydrogen atom, satisfying

$$(T + U - E_0)u = 0, \quad (T + V - E_0)v = 0, \qquad (2.2.2)$$

E_0 being the energy of the hydrogen ground state. However, we have no reason for preferring this assignment of the electrons to the one with 1 and 2 interchanged, $v(1)u(2)$. From the symmetry of the Hamiltonian we know that the exact solution must be symmetric or antisymmetric in the two electron positions (the symmetric orbital state belonging to opposite, the antisymmetric to parallel electron spins). We therefore choose, as approximate wave function,

$$\psi = u(1)v(2) \pm v(1)u(2). \qquad (2.2.3)$$

In this approximation we can estimate the energy as

$$\langle E \rangle = \frac{\langle \psi | H | \psi \rangle}{\langle \psi | \psi \rangle}. \qquad (2.2.4)$$

We insert our approximate wave function (2.2.3) in this estimate, and after a little algebra find

$$E - 2E_0 - \frac{e^2}{R} = \{\langle v|U|v \rangle + \langle u|V|u \rangle + \langle u(1)v(2)|W_{12}|u(1)v(2) \rangle$$
$$\pm [\Delta\langle u|U|v \rangle + \Delta\langle v|V|u \rangle + \langle u(2)v(1)|W_{12}|u(1)v(2) \rangle]\}/(1 \pm \Delta^2)$$

$$(2.2.5)$$

where Δ is the overlap integral

$$\Delta = \langle u|v \rangle. \qquad (2.2.6)$$

If the distance between the centers is greater than the diameter of a hydrogen atom, Δ is small compared to unity, and all the terms in the numerator of (2.2.5) are also small. The first three terms look as if they decrease only as R^{-1}, if R is the distance between centers, since, for example, the first term is the interaction of the Coulomb field of the first atom with the electron cloud of the second, and the third term is the Coulomb interaction between the two electron clouds. However, the sum of the three terms, plus the Coulomb interaction of the two protons, cancels at large distances, and the remainder decreases exponentially with R. The last three terms, which involve overlaps between wave functions centered on different atoms, also decrease exponentially.

We see, therefore, that the influence of the mutual disturbance of the two atoms is estimated, by this approximation, to be small, as long as R is large enough. It is therefore usually taken for granted that, for large enough R, the approximation is justified.

It is well known that indeed it makes very reasonable predictions about the interaction. The wave functions u and v may be taken as positive, and the Coulomb energies U and V are negative, whereas the electron-electron interaction W is positive. In practice, the negative terms dominate, and therefore E_+ is less than E_-. We find that the symmetric wave function, belonging to opposite spins, results in a net attraction, which makes the formation of a stable H_2 molecule possible, whereas the antisymmetric function, for parallel spins, gives repulsion. These facts form the basis of the quantum theory of the homopolar bond.

The smallness of the first perturbation term, however, is no guarantee that the perturbation series is well-convergent. This is demonstrated particularly clearly by noting that the relative position of the energies given by (2.2.5) depends on the choice of the interaction potentials. We may, with equal justification, apply the method to a system in which the repulsion between the electrons is not just the Coulomb force, but where there is a stronger interaction at short distances. In that way we can easily construct an example in which the last term in the numerator of (2.2.5) exceeds the sum of the two preceding ones, so that the square bracket becomes positive. In that case, E_- would be lower than E_+. However, this contradicts the correct answer, because there is a theorem by which the eigenfunction of a Schrödinger equation belonging to the lowest eigenvalue must be nodeless. Hence, the antisymmetric state, whose wave function changes sign on interchanging the electrons, and which therefore must have nodes, cannot be the ground state.

Let us recall the extremely simple derivation of this theorem. We use the variation principle in the form that the expectation value

$$\langle H \rangle = \frac{(\hbar^2/2m) \int dx (\text{grad } \phi)^2 + \int dx \, V(x)\phi(x)^2}{\int dx \phi(x)^2} \qquad (2.2.7)$$

for any function $\phi(x)$ which is differentiable almost everywhere must be higher than the ground state energy, and can become equal to the ground state energy only if ϕ is the ground-state wave function. Here x may represent any number of variables, and dx is the volume element in the space of these variables. Note that by expressing the kinetic energy in terms of the square of the gradient, rather than the integral of $-\phi\nabla^2\phi$, which is identical (by an integration by parts) provided ϕ has a second derivative almost everywhere, we have written a form in which the inequality is valid even if ϕ has a discontinuous derivative.

Assume now that the ground-state eigenfunction ψ_0 has nodes, i.e., that it is positive in some parts of space and negative in others. Then the function $\phi = |\psi|$ is not identical with ψ. If we insert this in the variation principle (2.2.7), we obtain the same result as with ψ, since the integrands are the same in both cases, except on the nodal surfaces, which are of measure zero. Thus the integral should equal the ground state energy, and this is ruled out by the variation principle, since ϕ is not identical with ψ (and, having a discontinuous derivative, could not be a solution of a Schrödinger equation).

The argument shows that the theorem is quite general, provided the potential is not spin-dependent (when the eigenfunctions of different symmetry would not be solutions of the same Schrödinger equation) or momentum-dependent (when the potential energy term would contain a double integration, containing $\phi(x)\phi(x')$, which would invalidate the comparison, as x and x' could belong to opposite signs of ψ).

We conclude that, for strong enough W_{12}, the Heitler-London expression (2.2.5) gives the wrong order of the levels, even though it becomes small for large R. There is no reason to believe that the answer is not a good approximation for two hydrogen atoms at a sufficiently large distance (the case to which the Heitler-London paper applies), but this has to be justified by more subtle arguments than relying just on the smallness of the overlap.

The surprising fact is that the accuracy of this classical method is much harder to establish than one is led to believe by a superficial inspection.

It is tempting to generalize our conclusion to the case of other molecules and to claim that for a diatomic molecule with two identical atoms the ground state should always have zero spin. This would set us well on the way to proving that ferromagnetism could not exist. However, such a generalization is not possible, because if we have more than one electron per atom, i.e., more than two electrons in all, the state of complete orbital symmetry is forbidden by the Pauli principle. It would have to be combined with a spin function that was totally antisymmetric, and since there are only two spin orientations, the function cannot be antisymmetric in more than two electrons at a time.

It remains true that the solution of the Schrödinger equation with complete orbital symmetry belongs to the lowest possible eigenvalue, but this is of no practical interest because this function does not belong to any physical state, and our theorem says nothing about the ordering of the less symmetric ones. So we need not be too greatly surprised by the existence of ferromagnetism.

2.3. GROUND STATE OF $2n$ FERMIONS IN ONE DIMENSION

In discussing the last item we noted that there are no general theorems guiding the ordering of the energy levels of more than two electrons. It is therefore surprising to find that in the particular case of particles moving in one dimension, one can still prove that, if the number of particles is even, the ground state still has spin zero. This surprising result was obtained by Lieb and Mattis, *Phys. Rev.*, 125, 164, 1962. We shall give a brief outline of their idea.

Consider $2n$ identical fermions in one dimension, confined, for definiteness, in a finite section, say $0 < x < L$. They are subject to an arbitrary interaction which does not depend on spin, and has a potential dependent only on the positions of the particles.

We may consider, without loss of generality, the situation in which the total spin component in the z direction vanishes, since this occurs for any resultant spin from 0 to n. Then n electrons must have positive s_z, and the rest negative. We can, further, assume that particles 1 to n have $s_z = +\frac{1}{2}$, and $n + 1$ to $2n$ have $-\frac{1}{2}$. Knowing the wave function for this situation, we can get it for any other distribution of the spins from the total antisymmetry.

The idea of Lieb and Mattis is now to consider the further restriction to that part of configuration space in which

$$0 \leq x_1 \leq x_2 \leq \cdots \leq x_n \leq L,$$
$$0 \leq x_{n+1} \leq x_{n+2} \leq \cdots \leq x_{2n} \leq L \qquad (2.3.1)$$

(no restriction is made about the relative position of any particle with $+$ spin to any of $-$ spin). If we know the wave function in the sub-space (2.3.1) we can immediately extend it to the rest of configuration space, since we can get there by permuting particles of each spin component with each other, and according to the Pauli principle the wave function is antisymmetric under such interchanges.

The variation principle can be written again in the form (2.2.7), but it is easy to see that for a trial function ϕ which obeys Pauli's principle it is sufficient to evaluate the integrals over the sub-space (2.3.1), since the sub-space obtained from this by permutation each give the same contribution, so that including them merely multiplies the numerator and the denominator by $(n!)^2$.

Within the sub-space we may choose ϕ arbitrarily, except that, to be able to continue it antisymmetrically and continuously, we must make ϕ vanish on all the boundaries of (2.3.1), in particular when the coordinates of any two particles with the same spin component coincide.

It is now clear that within this sub-space our theorem about the nodeless ground state applies. We know, therefore, that the eigenfunction of lowest energy has no nodes within the stated sub-space (the nodes required by the Pauli principle forming the borders of the sub-space). We also know that an eigenfunction belonging to resultant spin $S = 0$ is completely symmetric between the coordinates of all particles of opposite spin component, i.e., for any permutations within the sub-space, whereas a function belonging to $S \neq 0$ must contain further nodes.

(If the last statement does not seem obvious, remember that, for example, an eigenfunction with $S = 1$, $M^x = 0$ can be obtained by applying a spin rotation to the function for $S = 1$, $M_S = 1$, which has $n + 1$ up spins and $n - 1$ down spins. The latter must therefore be antisymmetric in $n + 1$ coordinates, and the spin rotation does not affect the orbital symmetry.)

This establishes the result stated at the beginning of this topic. It is very surprising that a result of such generality can be derived so simply. The result is, however, of only academic interest, because it is very essentially restricted to one dimension. This is true because it is only in one dimension that we can define a sub-space such that the rest of configuration space is obtained from it by permutations, and that its boundaries are formed by points at which two coordinates coincide, so that the wave function must vanish. In two or more dimensions the positions of two particles can be interchanged by moving them around each other, without ever coinciding. In going along such a contour the wave function must change sign if it is to be antisymmetric, and therefore must have a node somewhere, but it is impossible to predict where the node will be. Therefore the device used to separate the minimum number of nodes required by the Pauli principle from the additional ones works only in one dimension. This, too, is comforting in view of the existence of ferromagnetism.

3. STATISTICAL MECHANICS

3.1. PAULI PRINCIPLE IN METALS

The modern electron theory of metals started with the remark by Pauli that the exclusion principle, and hence Fermi-Dirac statistics, must be applied to all the electrons, in particular all the conduction electrons, in a piece of metal. This removed at once the difficulty about the paramagnetism of metals. According to classical statistics, the electron spins should be free to follow an applied magnetic field, and this would lead to a paramagnetic susceptibility following Curie's law, i.e., proportional to the inverse temperature, whereas in fact metals showed a paramagnetic susceptibility which was independent of temperature and much weaker than would correspond to Curie's law.

By Fermi-Dirac statistics, most electrons are in orbital states already containing two electrons of opposite spin. They cannot therefore align themselves with the external field without violating the exclusion principle. This is possible only for electrons in states of motion which are not completely filled, i.e., those with energies within a distance kT from the Fermi energy, E_F. Their number is less than the total number of electrons by a factor of the order of kT/E_F, and this factor explains both the temperature independence and the small value of the susceptibility.

This step opened the way to the solution of many other paradoxes of a similar kind. However, although the application of the Pauli principle to all the electrons in the system was clearly required by the basic rules of quantum mechanics, and confirmed by empirical knowledge, it left people with a rather uncomfortable feeling: If two electrons are at opposite ends of a metal wire of macroscopic dimensions, say a meter in length, is it not surprising that they can manage to avoid being in the same state of motion? How can each of them know what the other is doing?

The answer to this question is that it would indeed be difficult for two electrons located far from each other to affect each other's motion, but that this is also not required by the exclusion principle. In this law, "state of motion" comprises both position and momentum, as far as the uncertainty relation allows us to know them; if we can specify that the two electrons are in different regions of space, this is enough to ensure that they are in distinct states of motion, and thus to satisfy the Pauli principle. No further restriction on their motion is required.

To see this a little more clearly, let us idealize the problem by thinking of non-interacting free electrons in a one-dimensional "box" of length L. The simplest way of specifying the motion of the electrons is to choose stationary states which will have wave functions proportional to $\sin kx$, where the possible values of k are multiples of π/L. There can be at most two electrons of opposite spin in each such state, and if L is large, this can be expressed by saying there is a density L/π of electrons of each spin direction in k space.

For each of these electrons the position is then uncertain to within L; it may be anywhere within the box. In particular we may not assert that any two of these electrons could not happen to be close together. It is often convenient, however, to specify the location of the electrons to some degree. We then replace the sine functions by wave packets, such that a particular one may be centered at $x = na$ and be a modulated wave of wave number $k = mb$, n, m being integers. If these wave packets are to be orthogonal and complete, we must have $a \cdot b = \pi$. The stationary states mentioned before correspond to the special choice $a = L$, $b = \pi/L$. In general there will be L/a different locations, and at each the density of possible wave packets in k space is $1/b = a/\pi$, i.e., L/a times fewer than before.

The wave packets are not strictly stationary states; each contains a spread of kinetic energy, but to set up a distribution corresponding to statistical equilibrium we need to specify the energy only to an accuracy better than KT. Even for a temperature of 1°K, and a Fermi energy of 2 eV, the spread in the wave vector can still be a few times 10^3 cm^{-1}, so that electron positions can be specified to within 10μ or so without appreciably disturbing thermal equilibrium.

Any two of these wave packets, differing in the number n, which specifies location, or in m, which specifies momentum, or both, are orthogonal states, and therefore count, for the Pauli principle, as distinct states of motion.

In problems of the present kind, which involve counting states, it pays to avoid surprises on an altogether lower level, by keeping in mind the distinction between the "box" boundary condition we have used, and the cyclic boundary condition, which is often more convenient.

With the latter choice we require wave functions which repeat after a distance L. If these are written in the form exp(ikx), k must be a multiple of $2\pi/L$. The spacing of permissible k values is therefore twice what we had in the box, but we must remember that there was, for the box, no point in letting k become negative, whereas now exp(ikx) and exp($-ikx$) are distinct functions. For counting up the states in a given energy range, i.e., within a given range of k^2, the cyclic boundary condition gives us half the density of states, but spread over two equal intervals in k. Overall, both schemes therefore give the same answer if we apply each consistently.

3.2. IONIZATION

If atoms are in statistical equilibrium at a temperature T, in circumstances in which the interaction between atoms is negligible, the probability of an atom being in the state n is given by Boltzmann's expression

$$\frac{1}{Z} e^{-\beta E_n}, \tag{3.2.1}$$

where, as usual,

$$\beta = 1/KT, \tag{3.2.2}$$

K being Boltzmann's constant, and

$$Z = \sum_n e^{-\beta E_n}, \qquad (3.2.3)$$

the sum to extend over all states, and to include an integration over the continuous spectrum.

At low temperature, when one is interested mainly in states of low excitation, the relative probabilities of occupation can easily be read off from (3.2.1). However, even at low temperature, we cannot get absolute values, because the sum for Z in (3.2.3) diverges. This is true even for the contribution from the discrete spectrum. For example, in hydrogen there are, counting spin, $2n^2$ states of energy

$$E_n = -\frac{R}{n^2}, \quad n = 1, 2, 3, \ldots, \qquad (3.2.4)$$

where R is Rydberg's constant. These tend to 0 for large n, so the discrete sum for Z diverges as $\sum 2n^2$. The contribution of the continuous spectrum makes the divergence even worse.

We have to conclude, with some surprise, that the chance of the atom remaining in the ground state or in any other finite state is zero. To make this a little more quantitative, consider an atom in a finite volume, say a sphere of radius a. Then states extending over a radius much less than a will have the same energy as in free space, and states extending much beyond a will not exist. Since the mean radius is $a_0 n^2$, with a_0 the Bohr radius, we can find the right order of magnitude by cutting off the sum at $n = (a/a_0)^{1/2}$, which, for large a, gives $\frac{2}{3}(a/a_0)^{3/2}$. On the other hand, it is easy to see that the continuous spectrum contributes, in the same limit, an amount proportional to the volume, i.e., to a^3. So in a really large volume the dominant state for the electron is to be in a state of positive energy; the atom is ionized.

If we are really dealing with a single atom in an infinite volume, this is physically the correct answer, because the equilibrium for ionization depends on the available volume. This is evident from the consideration of the rate of ionization and recombination, which must balance in equilibrium. Whatever the mechanism, the rate of ionization is independent of the volume, whereas the rate of recombination depends on the chance of the electron meeting the nucleus again, which is inversely proportional to the volume.

In practical problems we are more often concerned with a large number of atoms in a large volume at finite, if perhaps low, density. In that case, we can use the eigenstates of the free atom only if their extension is less than the mean spacing between atoms, and we should, for each atom, not count states of a greater extension, since these overlap other atoms, and are then, in fact, approximated by the states of the other atoms, which will also be included. Similarly we must not attribute a continuous spectrum to each atom, since this would also involve double-counting. At low density, the continuous spectrum can be replaced by electrons moving in free space. Finally, we must allow for the mutual repulsion of the electrons, which means that the binding energy of an electron in the atom is substantially reduced if there is another electron in it already.

The full discussion, given, for example, very clearly in Landau and Lifshitz, *Statistical Physics*, §103, shows that, at temperatures at which the atom is not almost completely ionized, the excitation probability is negligible, so that one is concerned only with atoms in the ground state, ions (i.e., for hydrogen bare protons), and free electrons. The equilibrium can then be determined by the usual methods for finding the equilibrium in a reaction. The result given by Landau and Lifshitz takes, for hydrogen, the form that the degree of ionization, i.e., the ratio between the number of ions and that of all nuclei, is

$$\alpha = \left\{ 1 + \frac{P}{KT} \left(\frac{2\pi\hbar^2}{mKT} \right) e^{-\beta R} \right\}^{-1/2}, \qquad (3.2.5)$$

where P is the pressure, and m the electron mass.

We verify again that for infinite volume with a finite number of atoms, $P = 0$, the degree of ionization becomes complete at any temperature.

Historical note: The theory of thermal ionization was, I believe, first given by Saha, and the result used to be known as Saha's equation. It is still referred to as such by astrophysicists, but tends to be ignored in physics texts, with a few exceptions (such as the passage in Landau and Lifshitz referred to above). Physicists encountering this problem may therefore be tempted to fall into the trap which constituted our surprise here.

3.3. PERTURBATION THEORY FOR STATISTICAL EQUILIBRIUM

It is well known that all equilibrium properties of a system can be derived from the partition function Z, already defined in (3.2.3):

$$Z = \sum_n e^{-\beta E_n}, \quad \beta = \frac{1}{KT}. \tag{3.3.1}$$

In many cases, the energy eigenvalues E_n are not known exactly, and the Hamiltonian is of the form

$$H = H_0 + W \tag{3.3.2}$$

with only the eigenstates of H_0 known exactly. If W is small enough, we can find the eigenstates of H approximately by using a perturbation expansion, of which the first two terms are given by

$$E_n = E_n^0 + W_{nn} + \sum_{\substack{m \\ m \neq n}} \frac{|W_{nm}|^2}{E_n^0 - E_m^0}, \tag{3.3.3}$$

provided the unperturbed eigenvalues are all distinct. It is also known that, for the series to converge rapidly it is usually necessary that

$$|W| \ll \Delta E. \tag{3.3.4}$$

This is a somewhat symbolic condition. W is some suitable measure of the operator W, and ΔE is some typical value of the spacing between unperturbed energy levels. The precise limit must be determined in each problem by mathematical argument or by physical reasoning.

As an illustration, take the case in which one matrix element W_{mn} is greater than the distance between the energies of the unperturbed states it links. Then the mth term in the sum in (3.3.3) is also greater than $E_m^0 - E_n^0$, and, if there are no cancellations, the nth eigenvalue will be shifted by an amount exceeding its original distance from another one. This therefore means substantial changes in the spacings, and this will be felt in higher orders.

If we can accept the expansion, we can insert the perturbed eigenvalues in the expression (3.3.1) for the partition function, and reorder terms according to the powers of W which they contain. After some simple algebra, we find, to second order:

$$ Z = \sum_n e^{-\beta E_n^0} - \sum_n e^{-\beta E_n^0} W_{nn} - \frac{\beta}{2} \sum_{n,m} \frac{e^{-\beta E_n^0} - e^{-\beta E_m^0}}{E_n^0 - E_m^0} |W_{nm}|^2 + \cdots . $$

(3.3.5)

The sum in the last term does not exclude the term with $m = n$, since this just makes up the quadratic term due to the energy change of first order.

We note—and this is a pleasant surprise—that, unlike the perturbed energy, (3.3.3), the perturbed partition function does not show any sensitivity to small energy differences. If E_m^0 is very little different from E_n^0, the small denominator is offset by the small difference of the Boltzmann factors in the numerator.

It is easy to inspect the next term in the series and to verify that it has the same behavior. At this point we are encouraged to conjecture that the perturbation expansion of the partition function requires, for good convergence, not necessarily the condition (3.3.4), but instead the condition

$$ |W| \ll KT, $$

(3.3.6)

and this guess is indeed right. Physically, the situation is that W has the effect of moving each energy eigenvalue by an amount of the order of W. If this shift is greater than the level spacing, the determination of the exact position of the levels is a difficult problem. However, if the shift is less than KT, it does not change the value of the Boltzmann factor much, and one therefore does not need to know the precise position accurately.

The argument outlined here does not amount to a rigorous statement about the convergence of the series, any more than one can easily make rigorous general statements about the convergence of the perturbation series for the eigenvalues. This would in any case require a closer specification of the measure of W occurring in the conditions (3.3.4) or (3.3.6).

Historical note: The author initially encountered this pleasant surprise in a study of diamagnetism of conduction electrons, *Z. Physik*, 80, 763, 1933, when it seemed at first sight that the properties calculated from the discrete levels of the electron motion in a plane perpendicular to the magnetic field should be severely modified when the effect of collisions with phonons and impurities exceeded the small spacing of the magnetic levels.

3.4. Minimum Property of the Free Energy

The stability of the partition function which we discussed in the last topic, finds expression also in a minimum property of the free energy. We have already had occasion to refer to the variation principle for the ground state of a quantum system. For the excited states, individually, there does not exist an equally convenient bound (although we shall meet a very convenient relation in the next item), but it is again true that a change which raises one level tends to depress others, so that the partition function, and hence the free energy, shows great stability.

This property takes the following form: Let $u_1, u_2, u_3, \ldots$ be a set of orthogonal and normalized functions (not necessarily a complete set), and

$$H_{nn} = \langle u_n | H | u_n \rangle \qquad (3.4.1)$$

the corresponding diagonal elements of the Hamiltonian. Then the sum

$$\tilde{Z} = \sum_n e^{-\beta H_{nn}} \tag{3.4.2}$$

which would be the partition function if the H_{nn} were the correct eigenvalues, is at most equal to the true partition function

$$\tilde{Z} \leq Z \tag{3.4.3}$$

so that the "free energy" obtained from $\tilde{Z}$ is not less than the true free energy:

$$\tilde{F} = -KT \log \tilde{Z} \geq -KT \log Z = F. \tag{3.4.4}$$

In both relations the equality sign applies only if the set of functions u_n consists of all eigenfunctions of H.

This result was first given by the author in *Phys. Rev.*, 54, 918, 1934. I shall follow the proof outlined in that paper, though there are now more elegant proofs in the literature.

We want to show that the set of functions maximizing $\tilde{Z}$ must be the eigenfunctions of the Hamiltonian. It is not obvious a priori, however, that there is any set which does this, since it is not obvious that $\tilde{Z}$ has an upper bound. We therefore start by considering $\tilde{Z}_N$, the sum of the first N terms. This certainly has an upper bound $N \exp(-\beta E_1)$ if E_1 is the lowest eigenvalue of H, since no H_{nn} can be lower. There must therefore exist a set of N functions u_n maximizing $\tilde{Z}_N$. We shall show that these must be eigenfunctions of H, and then they must evidently be the first N, because replacing any of these by a higher one increases the appropriate energy and lowers $\tilde{Z}_N$.

We can always regard the u_n as the first N functions of a complete orthonormal set. If they are not eigenfunctions of H, there must exist for at least one of them, say u_n, at least one non-vanishing off-diagonal element, say H_{nm}, with $n \leq N$. m may or may not exceed N. If $m > N$, we can replace u_n by $u_n + \lambda u_m$, with infinitesimal λ. This changes H_{nn} by $\lambda H_{nm} + \lambda^* H_{mn}$, neglecting terms of second order in λ, while not affecting any of the other diagonal elements. By suitable choice of the phase of λ this change can be made negative, so that H_{nn} is reduced, and $\tilde{Z}_N$ increased; the set of functions did not maximize $\tilde{Z}_N$.

Alternatively, m may belong to one of the first N functions. In that case we replace u_n and u_m by

$$u_n + \lambda u_m, \quad u_m - \lambda^* u_n \tag{3.4.5}$$

again with infinitesimal λ, preserving the orthonormality to first order in λ. This changes H_{nn} and H_{mm} into

$$H_{nn} + \Delta, \quad H_{mm} - \Delta \tag{3.4.6}$$

where

$$\Delta = \lambda H_{nm} + \lambda^* H_{mn}. \tag{3.4.7}$$

(The sum of the two diagonal elements is unchanged, as it should be.) Hence $\tilde{Z}_N$ changes, to first order, by

$$-\beta \Delta (e^{-\beta H_{nn}} - e^{-\beta H_{mm}}). \tag{3.4.8}$$

If the two diagonal elements are different, the quantity in the parentheses is non-zero; hence, for a suitable sign of Δ, the whole term can be made positive, and this can be achieved by a suitable phase of λ, according to (3.4.7). In this case the u_n could also not have maximized $\tilde{Z}_N$.

This leaves the case in which $H_{nn} = H_{mm}$. Since the first-order change in $\tilde{Z}_N$ now vanishes, we have to consider the second-order term. It is true, even to second order, that the changes in H_{nn} and H_{mm} are opposite and equal, since their sum is the change in the trace of a 2×2 matrix under a unitary transformation, and therefore the second-order change in $\tilde{Z}_N$ is

$$\frac{1}{2} \Delta^2 \left[\left(\frac{d^2 e^{-\beta E}}{dE^2} \right)_{H_{nn}} + \left(\frac{d^2 e^{-\beta E}}{dE^2} \right)_{H_{mm}} \right] = \beta^2 \Delta^2 e^{-\beta H_{nn}}. \tag{3.4.9}$$

This is always positive. It follows that there is no case in which the u_n can maximize $\tilde{Z}_N$, unless they are eigenfunctions of H.

We have now proved the theorem for any finite N, and it must therefore hold also in the limit of infinite N. If it is true for a complete orthonormal set, it is true a fortiori for an incomplete set, since the latter can be converted into a complete set by adding further functions, which further increases $\tilde{Z}$.

The proof has been written out for the Boltzmann distribution $e^{-\beta E}$, but it is equally valid for any function with a negative first, and positive second, derivative. It can therefore also be used for dealing with the free energy of a system of non-interacting bosons or fermions in terms of a one-particle Hamiltonian.

The practical use of this minimum property is limited by the fact that in applications one is usually concerned with derivatives of the free energy with respect to parameters, or with differences. For these there do not exist general inequalities. One can, however, use it to select the best of a family of sets of trial functions, by minimizing the free energy through a choice of the parameters distinguishing different sets within the family, just as one often uses the minimum property of the ground state, which is the limit of the present theorem for zero temperature.

3.5. VARIATION PRINCIPLE FOR FIRST N STATES

If we are interested, not in the statistical equilibrium, but in the position of individual excited energy levels, the theorem discussed in the last item provides no help. There exists an inequality based on the usual variation principle, which states that the expectation value

$$\langle E \rangle = \frac{\langle \phi | H | \phi \rangle}{\langle \phi | \phi \rangle} \tag{3.5.1}$$

cannot lie below the nth eigenvalue E_n, if the trial function ϕ is orthogonal to the first $n - 1$ eigenfunctions:

$$\langle E \rangle \geq E_n, \quad \text{if} \quad \langle \psi_m | \phi \rangle = 0, \quad m = 1, 2, \ldots, n - 1. \tag{3.5.2}$$

This inequality cannot be used in general without a knowledge of the first $n - 1$ eigenfunctions; replacing the exact ψ_m by approximate eigenfunctions, for example those obtained variationally, can lead to substantial errors. An important exception is the case in which the Hamiltonian possesses a known symmetry, and we are looking for the lowest state of a given representation of that symmetry. Since functions of different symmetry are automatically orthogonal, choosing the right symmetry property for ϕ will ensure the orthogonality required in (3.5.2).

A more useful way of placing lower bounds on excited levels is provided by the following theorem: Let $u_1, u_2, \ldots, u_N$ be a set of orthonormal functions, and

$$H_{nm} = \langle u_n|H|u_m \rangle \tag{3.5.3}$$

the matrix elements of the Hamiltonian between them. Let W_s be the eigenvalues of the $N \times N$ matrix thus defined, ordered in sequence so that

$$W \leq W_2 \leq W_3 \leq \cdots \leq W_N.$$

Then each W_s is an upper bound to the sth eigenvalue of the full Hamiltonian:

$$E_1 \leq W_1, E_2 \leq W_2, \ldots, E_N \leq W_N. \tag{3.5.4}$$

It came as a surprise to me to learn of the existence of this theorem a few years ago from Dr. D. M. Brink. It can indeed be found in the mathematical literature, but the proofs which I found there were quite sophisticated and not easy to follow. It was therefore another surprise to find that it can be proved in a very simple and transparent way.

Consider first the case $N = 2$. We define a quadratic form in two complex variables a_1, a_2:

$$h = \frac{\sum\limits_{n,m=1}^{2} H_{nm} a_n{}^* a_m}{\sum\limits_{n=1}^{2} a^*{}_n a_n}. \tag{3.5.5}$$

Evidently h is equal to the expectation value of the Hamiltonian for a trial function

$$\phi = \sum_{n=1}^{2} a_n u_n. \tag{3.5.6}$$

It is also a well-known theorem in algebra that the values which h can take for any choice of the a_n cover the range

$$W_1 \le h \le W_2. \tag{3.5.7}$$

This shows, in the first place, that I can find a function ϕ which makes the variation quantity (3.5.1) equal to W_1, and by the usual variation principle this must be above the ground-state energy E_1. Thus the first part of (3.5.4) is essentially just an expression of the usual variation principle.

To find an upper bound for E_2 we note that among the trial functions (3.5.6) we can find one orthogonal to the ground-state eigenfunction. Indeed the condition

$$\langle \psi_1 | \phi \rangle = a_1 \langle \psi_1 | u_1 \rangle + a_2 \langle \psi_1 | u_2 \rangle = 0 \tag{3.5.8}$$

certainly can be solved by a suitable choice of a_1, a_2, not both vanishing. The expectation value of the Hamiltonian for this function must not lie below E_2 because of the generalized variation principle (3.5.2), yet it cannot lie above W_2 because of (3.5.7). This proves our theorem for $N = 2$.

Next consider $N = 3$. By exactly the same reasoning as for $N = 2$ we can see immediately that W_1 and W_3 are upper bounds for E_1 and E_3, respectively, but the inequality for the middle state is not immediately obvious. To cover this, make a unitary transformation of the u_n which diagonalizes the 3×3 matrix H_{mn}. This does not change the eigenvalues of the matrix. In terms of the new functions $\tilde{u}_1, \tilde{u}_2, \tilde{u}_3$, H is diagonal and has the eigenvalues W_1, W_2, W_3 as diagonal elements. We now discard u_3, and are left with a 2×2 matrix with eigenvalues W_1 and W_2. From the theorem for $N = 2$, which we have proved, we deduce the missing relation

$$E_2 \leq W_2.$$

It is now clear that we can proceed to any N by induction.

Practical applications are again limited by the fact that we are often most interested in the energy differences, on which the theorem provides no direct information.

3.6. INFLUENCE OF BOUNDARY CONDITIONS

In dealing with the statistical mechanics of large uniform systems in space, such as electrons in metals, or lattice vibrations in crystals, etc., it is often convenient to replace the actual physical boundary condition, which may be complicated and sensitive to the physical state of the surface, by a simpler one, such as the "cyclic boundary condition"

$$\psi(x + L_1, y, z) = \psi(x, y + L_2, z) = \psi(x, y, z + L_3) = \psi(x, y, z)$$

for the one-particle wave function ψ, or for whatever other quantities describe the system. This gives us the right statistical weights for a body of linear dimensions L_1, L_2, L_3 and is much easier to handle. Provided the linear dimensions are very large on an atomic scale, we accept intuitively that this will cause no serious error in computing bulk quantities.

This assumption is sometimes queried, and it is therefore useful to have a simple argument justifying it, which will also give us information about the circumstances in which the approximation may prove inadequate.

I shall present the argument for the case of one-particle eigen-functions for definiteness, but with an appropriate change of notation it applies equally to other systems. For the one-particle problem we then assume a macroscopically uniform system; there could therefore be a periodic potential, or an irregular potential which is uniform on the average.

The important quantity for treating the statistical mechanics of this problem is the density of states in energy, given by

$$\rho(E) = \sum_n \delta(E - E_n). \tag{3.6.1}$$

Full knowledge of $\rho(E)$ is identical with an accurate knowledge of all eigenvalues. Their position is certainly sensitive to the boundary conditions.

This represents more accurate knowledge than we require, however, because in evaluating the partition function or other equilibrium properties, we are concerned with integrals in which the density of states is multiplied by a function of energy which varies appreciably only over energy intervals of the order of KT. It is therefore sufficient to know an average density,

$$\tilde{\rho}(E) = \int_{-\infty}^{\infty} D(s)\rho(E + s) \, ds. \tag{3.6.2}$$

The weight factor D used in the averaging should satisfy

$$\int D(s) \, ds = 1. \tag{3.6.3}$$

and its range Δ should be small compared to KT:

$$\Delta^2 = \int D(s)s^2 \, ds \ll KT. \tag{3.6.4}$$

For a large system this may still cover many eigenvalues.

It is convenient to express $\rho(E)$ as a space integral

$$\rho(E) = \int d^3r\, \rho(\mathbf{r}, E), \tag{3.6.5}$$

where

$$\rho(\mathbf{r}, E) = \frac{1}{2\pi} \int_{-\infty}^{\infty} dt\, e^{iEt/\hbar} G(\mathbf{r}, \mathbf{r}, t). \tag{3.6.6}$$

Here G is the Green function

$$G(\mathbf{r}, \mathbf{r}', t) = \sum_n u_n(\mathbf{r}) u_n(\mathbf{r}')\, e^{-iE_n t/\hbar} \tag{3.6.7}$$

in terms of the energy eigenfunctions. One easily verifies that the expression for $\rho(E)$ is equivalent with the definition (3.6.1). We have introduced G, because it is also the function giving the time evolution of the wave function:

$$\psi(\mathbf{r}, t) = \int G(\mathbf{r}, \mathbf{r}', t) \psi(\mathbf{r}', 0)\, d^3r'. \tag{3.6.8}$$

The averaging with the weight D can also be done for the local density of states, leading to

$$\tilde{\rho}(\mathbf{r}, E) = \frac{1}{2\pi} \int ds\, D(s) \int dt\, e^{i(E+s)t/\hbar} G(\mathbf{r}, \mathbf{r}, t)$$

$$= \int dt\, f(t)\, e^{iEt/\hbar} G(\mathbf{r}, \mathbf{r}, t), \tag{3.6.9}$$

where

$$f(t) = \frac{1}{2\pi} \int ds\, D(s)\, e^{ist/\hbar} \tag{3.6.10}$$

is the Fourier transform of D. Now since D is a smooth weighting function of spread Δ, its Fourier transform need be appreciable only for time of the order

$$\tau = \hbar/\Delta. \qquad (3.6.11)$$

This shows that the local density of states, $\tilde{\rho}(\mathbf{r}, E)$ given by (3.6.9) depends only on the time evolution function G for times less than τ. If $\mathbf{r}$ is a point distant R from the nearest surface, and v a typical velocity of propagation for the system under consideration, the effect of a change in ψ at $\mathbf{r}$ on the value of ψ a time t later can depend on the nature of the surface only if there has been time for a disturbance to travel from $\mathbf{r}$ to the surface and back, i.e., if $t > 2R/v$. It follows that $\tilde{\rho}(\mathbf{r}, E)$ can be sensitive to the boundary condition used only for a layer of thickness $\frac{1}{2}v\tau$ inside the boundary of the medium. If this thickness is small compared to the linear dimensions of the system, the correction is small and proportional to the surface area, as expected for a surface effect.

Since Δ is restricted only by (3.6.4) to be smaller than KT, τ need only be larger than $\hbar/KT$. At room temperature, this limit is about 2×10^{-14} sec. For electrons in metals a typical value for v would be the velocity of an electron with the Fermi energy, which is of the order of 10^8 cm/sec. This means that R should be large only compared to 2×10^{-6} cm, and for most practical problems this is negligibly small. For lattice vibrations the relevant velocity would be the velocity of sound, typically of the order of 10^5 cm/sec, which makes the limit on R even smaller.

For all systems we can be sure that no influence can propagate faster than light, so for room temperature the limit on R can be at most 6×10^{-4} cm.

For very small systems, particularly at low temperatures, the surface corrections may no longer be negligible; the considerations sketched above may then help to identify situations for which this is the case.

The surprising aspect of this problem was, for me, first that there were eminent physicists who needed to be convinced that the equilibrium properties of a large system would not be affected by the use of an unrealistic boundary condition. A further surprise was to see how easy it was to give a physically transparent proof of the statement. I published an argument along these lines, applied to the equation of state of a relativistic gas (*Monthly Notices, R. Astron. Soc.*, 96, 780, 1936), when Eddington had queried the usual result for that case, and a proof for the lattice specific heat of a solid (*Proc. Nat. Inst. India*, 20, 121, 1954) when Raman disputed the use of a cyclic boundary condition for that problem.

3.7. SPECIFICATION OF SURFACE ENERGY

The discussion of the previous topic indicates that for a reasonably large volume, the surface energy is a local quantity which can be discussed by considering each surface element separately. In general, the radius of curvature of the surface is of the order of the linear dimensions of the system, and therefore by assumption large compared to the surface thickness. One may then treat each surface element as plane, and attribute to it the surface energy, per unit area, of an infinite plane surface bounding a half-space.

Such approaches are used, for example, in nuclear theory, where the surface energy of a large nucleus contributes a term to the total energy which can be identified by fitting a semi-empirical formula to the masses of nuclei. The theory of this surface energy depends of course on the assumptions made about the nucleon-nucleon interaction, and on the approximations used for the nuclear many-body problem. The earliest approaches used a "well-behaved" force without singular repulsive core for the former, and an independent-particle model for the latter, and the point to be made arises simply and clearly in a discussion at this level, although we would not today regard such a calculation as realistic.

An apparent objection to the ideas outlined above comes from the fact that, in a finite volume, the particle states are discrete, and the total distribution of particles is a sum over these discrete states, rather than an integral over a continuous distribution, as it would be for an infinite volume. One of the differences between a finite and an infinite volume comes from the difference between the sum and the integral. It turns out that this difference is, relative to the integral, proportional to the reciprocal of the linear dimensions, like the ratio of surface to volume. This consideration, first pointed out, I believe, by Feenberg, *Phys. Rev.*, 60, 204, 1941, would make us suspect that the calculation of the surface energy is much more difficult than I have indicated, because we cannot discuss the spacing of the quantized energy levels without specifying the shape of the volume. In addition, an estimate by Feenberg of the magnitude of this quantization effect suggested that it was much larger than the empirical surface energy.

It was therefore an agreeable surprise when W. J. Swiatecki (*Proc. Phys. Soc.*, A64, 226, 1951) found that in a correct statement of the problem the quantization term does not enter, and that the simple approach I have sketched is perfectly in order. To understand this, let us consider, for simplicity, a one-dimensional problem, and ignore spin and isotopic spin.

We assume therefore N identical fermions moving freely in one dimension in the "box"

$$0 < x < L. \tag{3.7.1}$$

In the limit of large N and large L, but with a finite "density" N/L, we then expect the leading term in the energy, which is the analogue of the volume energy, to be proportional to N. The next term, corresponding to the surface term, is here an "end effect", independent of N. Similarly the kinetic and potential energies, taken separately, will have a leading term proportional to N, and an end correction independent of N, followed possibly by terms behaving in the limit like negative powers of N, in which we are not interested.

It is easy to evaluate the kinetic energy in the independent-particle approximation. The eigenfunctions are

$$\psi_n = \left(\frac{2}{L}\right)^{1/2} \sin \frac{\pi n x}{L}, \tag{3.7.2}$$

where n is a positive integer. The states from $n = 1$ to N will be occupied. The energy of the nth state is $(\hbar^2/2m)(n\pi/L)^2$, and the total kinetic energy is

$$E_{\text{kin}} = \frac{\hbar^2}{2m}\left(\frac{\pi}{L}\right)^2 \sum_{n=1}^{N} n^2 = \frac{\hbar^2}{2m}\left(\frac{\pi}{L}\right)^2 \left(\frac{1}{3}N^3 + \frac{1}{2}N^2 + \frac{1}{6}N\right). \tag{3.7.3}$$

This can also be expressed as

$$E_{\text{kin}} = \frac{\hbar^2 \pi^2}{2m}\left(\frac{N}{L}\right)^2 \left(\frac{1}{3}N + \frac{1}{2} + \frac{1}{6N}\right), \tag{3.7.4}$$

which shows the expected behavior; the second term in the parentheses appears to be the end term. This suggests for each end a contribution of $\frac{1}{4}(\hbar^2\pi^2/2m)(N/L)^2$.

However, this leaves out of account the behavior of the particle density. From the wave functions (3.7.2) it is easy to obtain the particle density, which is

$$\rho(x) = \frac{N + \frac{1}{2}}{L} - \frac{\sin 2\pi(N + \frac{1}{2})x/L}{2L \sin \pi x/L}. \tag{3.7.5}$$

This has the form sketched in Figure 3.1. It must of course vanish at $x = 0$ or L, since all the wave functions vanish there, and in the interior, far from the ends, it is more or less constant. This constant value is $(N + \frac{1}{2})/L$. This is natural, since there is a nearly empty space at each end of the box, of a width of about $L/4N$. In other words, the bulk of the distribution we have studied is at a density $(N + \frac{1}{2})/L$. Since the kinetic energy of N particles in one dimension at density ρ can be expressed as $(\hbar^2\pi^2/6m)N\rho^2$, we see that the increase in density accounts for an increase in the kinetic energy

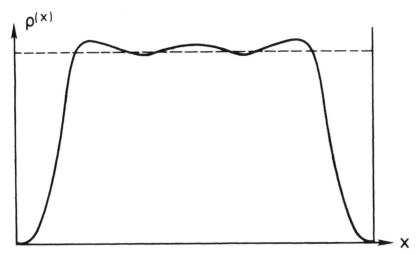

Figure 3.1   Particle density for a one-dimensional Fermi gas in a box.

of $\hbar^2\pi^2 N^2/6mL^2$, which is two-thirds of the second term in (3.7.4).

The remainder, or one-third of the apparent end effect, should then be a genuine end term if we compare systems of given internal density, and should agree with what we find by looking at one end of an infinitely long one-dimensional box. To verify this, we note that the density of kinetic energy with the eigenfunctions (3.7.2) is:

$$t(x) = \sum_n \left(-\frac{\hbar^2}{2m}\right)\psi_n\frac{d^2\psi^n}{dx^2} = \frac{\pi^2\hbar^2}{L^3}\sum n^2 \sin^2\frac{n\pi x}{L}.$$

In the limit of infinite L we can replace the sum by an integral:

$$t(x) = \frac{\hbar^2}{m\pi}\int_0^K k^2 dk\,\sin kx, \quad k = \frac{n\pi}{L}, \quad K = \frac{N\pi}{L}$$

$$= \frac{\hbar^2}{6m}K^3\left[1 - \frac{3\sin 2Kx}{2Kx} - \frac{6\cos 2Kx}{(2Kx)^2} + \frac{6\sin 2Kx}{(2Kx)^3}\right]. \quad (3.7.6)$$

We compare the total kinetic energy of this with that of the same number of particles at density $\rho_0 = N/L$, when the kinetic energy density would be $t_0 = (\hbar^2 K^2/6m)\rho_0$. The excess kinetic energy is

$$\int dx \left[t(x) - \frac{t_0}{\rho_0} \rho(x) \right] = \frac{\hbar^2 K^2}{24M}, \qquad (3.7.7)$$

which is the same as the extra term for each end left over after allowing for the density change.

The calculation for three dimensions is a little longer, but proceeds in the same way and with the same conclusion. This justifies the calculation of the surface energy by considering a semi-infinite system. In addition to the kinetic energy, one must of course also work out the potential energy contribution to the surface energy. In the independent particle model, which forms the basis of our discussion, this can be obtained from the knowledge of the single-particle wave functions, and the two-body interactions.

The presence of the surface energy, which also causes a surface tension, will cause the density inside a finite body to be slightly larger than in the limit of infinite size, and at first sight we may think that this calls for another correction. However, if we start from a large body in equilibrium, the density will be such as to minimize the energy; the increase in energy caused by a small change in the density is therefore proportional to the square of the density change, and therefore to the square of the surface tension. This is negligible for our purpose. The importance of the density change in Swiatecki's argument is due to the fact that it relates to the surface contribution of the kinetic energy. The kinetic and potential energies separately are not minima at the bulk density. If one had applied the calculation based on the quantized states to both kinetic and potential energy, the implied change of the interior energy density would have canceled out, to the first order in $1/N$, relative to the volume energy. The surface contribution to the potential energy is, however, more difficult to calculate, and one is therefore tempted to rely on the fact that it should be positive, since the main effect of the surface is that some of the particles are lacking neighbors, and the net attraction due to the neighbors. If this were true, the kinetic-energy contribution would be a lower bound to the total surface energy. This is the point where the change in interior density comes in: At the equilibrium density the kinetic energy is an increasing function of density. Since the total energy is a minimum, the potential energy decreases (i.e., becomes more negative) with increasing density. By using a model in which the interior density has been increased, we have therefore made the potential-energy contribution to the surface energy negative; the kinetic-energy contribution is no longer a lower bound.

These considerations clear the way for practical calculations of surface energy. If one uses the approximation of independent particles, one has to generate a suitable model for the particle states near the surface of a semi-infinite volume, most conveniently in terms of the eigenfunctions in a suitable potential, and then evaluate the density of kinetic energy and of the two-body interaction potentials near the surface. In principle one should then vary the assumed potential until the surface energy is minimized. The first work along these lines was carried out in the paper by Swiatecki to which I have referred. By now there are many studies available which go beyond the independent-particle approximation.

The procedure at which we have arrived is probably just what we would have done if we had not thought much about possible complications and possible surprises. Thinking a little more deeply, but not deeply enough, about the problem, one is taken aback by the need to look at the spectrum of quantized levels. It takes further insight to realize that this concern is not warranted, and that the "naive" approach is quite in order. This situation is by no means unique.

3.8. IRREVERSIBILITY

We turn next to one of the most fundamental questions of statistical mechanics, to which the answer has been known to some for a long time, but does not appear to be known very widely even today. The question is about the precise origin of the irreversibility in statistical mechanics.

We know that statistical mechanics is based on the equations of

Newtonian mechanics and, where appropriate, Maxwell's equations for the electromagnetic field. All these laws are reversible in time in the sense that for any motion permitted by these laws, with the appropriate electromagnetic fields, the one obtained by reversing the time direction, i.e., replacing t by $-t$, is also a possible solution. In constructing this reversed motion, one must of course reverse the sign of all velocities, and with it the sign of the magnetic field. In other words, the laws of mechanics and of electromagnetism do not make any distinction between past and future.

Yet we also know that statistical mechanics leads to all the known laws of thermodynamics, including the Second Law which states that for an isolated system the entropy can never decrease:

$$\frac{dS}{dt} \geq 0. \tag{3.8.1}$$

This law is certainly not symmetric in time; if we interchanged past and future the entropy would tend to decrease. How did we get, from basic reversible equations, to a manifestly irreversible result?

The same surprise also occurs in quantum mechanics. The time-dependent Schrödinger equation, which then represents our mechanical law, admits a solution in which the time is reversed, provided we change the wave function ψ to its complex conjugate, which has the effect of reversing all velocities. The language of quantum mechanics is necessarily more complicated than that of classical physics. We shall therefore confine our discussion to classical statistical mechanics, in which the paradox is very clearly seen.

Let us consider a typical irreversible situation. We choose a box divided into two chambers by a wall with a communicating window which can be opened and closed. Initially, at time $t = 0$, the gas pressures in the two chambers differ, one chamber containing N_1, the other N_2 molecules of the same gas.

If all other variables are kept fixed, we can take the entropy as a measure of the distribution of gas over the two chambers, being a maximum when the densities are equal. (The actual relation is

$S = \text{const.} - K(N_1 \log N_1 + N_2 \log N_2)$, but we do not have to use this.) If we then evaluate the entropy by using the laws of mechanics, with the appropriate randomness assumptions of statistical mechanics, we obtain a result looking like the solid curve in Figure 3.2, in line with the Second Law.

Some textbooks explain this paradox by saying that, whereas particle mechanics makes predictions about the motion of individual particles, statistical mechanics makes probability statements about large ensembles of particles. This is true, but it does not explain why the use of probabilities and statistics should create a difference between past and future where none existed before.

The real answer is quite different. Suppose from $t = 0$ when we assumed the particles distributed at random within each container and to move in random directions, we follow the particle trajectories, not for positive times, but for negative t, i.e., into the past. This will give a curve for the entropy looking like the broken curve in Figure 3.2, and it will be the mirror image of the solid curve.

We see therefore that the symmetry in time is preserved fully in these two calculations. However, the solid curve to the right describes a situation which occurs in practice, and therefore provides

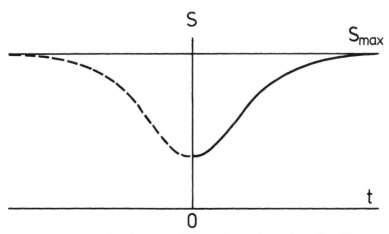

Fig. 3.2   Entropy vs. time for gas in a divided enclosure, given an inequality of density at $t = 0$. Full curve: actual behavior. Broken curve: hypothetical behavior in the past.

the answer to a realistic question, whereas the broken curve to the left does not.

The situation to which the broken, left-hand curve would be applicable would be the following: Arrange for particles at $t = 0$ to be distributed in given numbers over the two chambers, their positions being random in each chamber, and their velocities having a Maxwell-Boltzmann distribution. Ensure that prior to $t = 0$, at least after some finite time $-T$, there was no external interference, and observe the state of affairs at $t = -T$. This is evidently impossible; the only way in which we can influence the distribution of molecules at $t = 0$ is by taking action prior to that time.

We thus see that the asymmetry arises, not from the laws governing the motion, but from the boundary conditions we impose to specify our question. The situations with which we are concerned in practice are described by *initial* conditions, corresponding to the right-hand part of Figure 3.2. The left-hand part would correspond to *terminal* conditions, which are what we obtain from initial conditions upon reversing the time, and these have no practical relevance.

This situation is not substantially changed if we consider a system containing only small numbers of molecules, rather than the macroscopic situation envisaged above. In that case, fluctuations are not negligible, and for a given state of affairs at $t = 0$, the system can develop in time in a number of different ways. It is possible, for example, that for $N_1 < N_2$ some of the few molecules in chamber 1 may happen to be close to the window, and moving towards it. In that case, N_1 will initially decrease further, and N_2 increase accordingly, so we may see a temporary decrease of the entropy. The right-hand side of Figure 3.3 illustrates a few of the possible curves we may expect in this case. Each individual curve will in general be unsymmetric if continued to negative times, but its mirror image is another possible, and equally probable, behavior. Again only the right-hand half of the figure represents practical situations with an arranged state of affairs at $t = 0$.

There is, however, one difference between the system with only

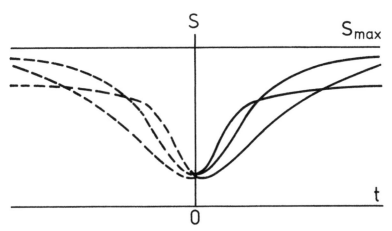

Figure 3.3   As Figure 3.2, but for a small number of molecules, so that fluctuations are not small.

a few molecules and the macroscopic one: When a system is left to itself, we know from the general results of statistical mechanics that it will spend most of its time near the equilibrium value of the parameters, i.e., near maximum entropy, but will show occasional fluctuations with lower entropy. If we are confronted at $t = 0$ with our two-chamber system having different densities in the chambers, with the window open, and knowledge that at least for a preceding period of length T there had been no interference, then it would be a reasonable conjecture for the small system that this was due to a fluctuation, and in that case we would have to describe the system by one of the curves in Figure 3.3 using both sides. We could here allow a curve which dips below the fixed $t = 0$ point either for positive or negative time, but the probability of a fluctuation decreases rapidly with its amplitude, so that the number of curves to choose from is very much less if we want them to dip appreciably below the $t = 0$ value than if we do not. In this interpretation both sides of the figure are applicable, and in this case there is also no irreversibility in the conclusions.

For the macroscopic system this interpretation is not open to us. For a liter of air at normal temperature and pressure, a fluctuation by which the density in one-half of the container differed from the other by even a percent will not occur even over an astronomical time. To attribute a substantial difference to a fluctuation would therefore be entirely unreasonable. In that case a much more reasonable hypothesis would be that, prior to the time $-T$, after which we know the system to have been undisturbed, someone opened the window, and that at that time the densities were even more different. In that case we would draw a curve like that of Figure 3.4, which is indeed unsymmetric in time. But it is also clear that, besides statistical mechanics, we have also used some prejudice about what human interference might have been responsible for the system being in the non-equilibrium state at $t = 0$.

We may try to analyze the problem somewhat more deeply by asking why it is that we can easily perform experiments in which initial conditions have to be specified, but never any requiring terminal conditions. This is the real distinction between past and future. A little thought shows that this is connected with the fact that we can remember the past, and that we can make plans for the future, but not vice versa. It is evident that these statements are correct, but they do not follow from any known law of physics.

We could speculate that these facts have something to do with

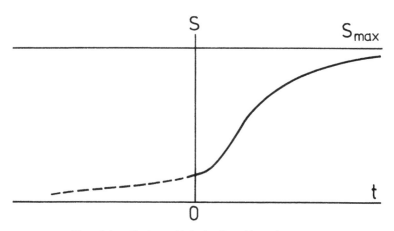

Figure 3.4   Conjectured behavior if past history is unknown.

the way our brain functions, but we have no way of explaining the origin of this one-sidedness. We must not, of course, try to attribute it to some thermodynamic irreversibility in our brain cells, because this would amount to invoking the Second Law in a study aimed at understanding the origin of the Second Law, and we would have a circular argument.

Not all studies of irreversible behavior contain the elements of initial conditions in as manifest a manner as the example we have discussed. Boltzmann's famous "H theorem", which is the statistical-mechanics version of the Second Law, was proved by considering the collisions of molecules in a dilute gas without introducing some specific initial instant. It turns out, however, that Boltzmann's Stosszahl-Ansatz implies an equivalent assumption, and that this is responsible for the irreversibility.

To see this, we shall limit ourselves to the so-called "Lorentz gas," i.e., to non-interacting molecules scattered by fixed scatterers. We require to know in a non-equilibrium situation, for example for a drift motion, the number of molecules scattered in a time interval dt from an initial state of motion a, specified by direction of motion and velocity and, if necessary, the values of any internal parameters, to a final state b. For a particular scatterer there will be a differential cross section σ_{ab} for this event. If we draw, as in Figure 3.5, a cylinder of base area σ_{ab} and length $v\,dt$, aimed at the scatterer in the direction of the initial velocity v_a, then the number of molecules scattered in time dt into the state b will equal the number of molecules contained initially in this cylinder. We may therefore write the number scattered in the specified manner in the form:

$$\rho_a \sigma_{ab} v_a\,dt, \qquad\qquad (3.8.2)$$

where ρ_a is the density of molecules of type a in the cylinder indicated. The Stosszahl-Ansatz of Boltzmann now consists in the seemingly innocuous assumption that ρ_a equals the average density of molecules of this type anywhere in the gas, i.e., that there is nothing exceptional about the particular cylinder we have defined.

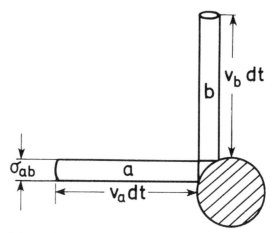

Figure 3.5   Boltzmann's Stosszahl-Ansatz. Cylinder *a* contains the molecules which, with the specified velocity, will collide with the target in thè time interval *dt* and be scattered into *b*. Cylinder *b* contains those scattered from *a* in dt. The Ansatz assumes that the density of such molecules in *a* is the same as elsewhere in the gas, which has the consequence that the density in *b* is not the same.

This assumption is the origin of irreversibility, because if it is true, the corresponding statement about the cylinder labeled *b* in the figure is *not* true. The only special thing about cylinder *a* is that it contains the molecules which are going to collide with the scatterer; cylinder *b* contains those which have just collided. In non-equilibrium conditions, for example in the presence of a drift motion in the *a* direction, there will be more molecules in the gas as a whole moving in the *a* direction than in the *b* direction. Scattering by the center will therefore tend to increase the number in the *b* direction. If ρ_a is the same as elsewhere in the gas, ρ_b must then be greater than the average.

If the scattering is compared to the time-reversed situation, we see a difference. To reverse the direction of time, we have to replace each molecule in *b* by one of the opposite velocity, and have them scattered by the target to travel in the direction opposite to that of *a*. The number would not be changed, and if ρ_b in the cylinder *b* differs from the average over the whole gas, it will also differ from what, with Boltzmann, we should assume about the inverse process.

It seems intuitively obvious that the molecules should not be influenced by the fact that they are going to collide, and very natural that they should be affected by the fact that they have just collided. But these assumptions, which cause the irreversibility, are not self-evident. If we assume, however, that the state of the gas was prepared in some manner in the past, and that we are watching its subsequent time development, then it follows that correlations between molecules and scattering centers will arise only from past, but not from future, collisions. This shows that the situation is, in principle, still the same as in our two-chamber problem.

We have recognized the origin of the irreversibility in the questions we ask of statistical mechanics, and we have seen that their lack of symmetry originates in the limitations of the experiments we can perform. The "arrow of time" appears to be in our minds. As long as we have no clear explanation for this limitation, we might speculate whether the time direction is necessarily universal, or whether we could imagine intelligent beings whose time runs opposite to ours so that, from our point of view, they could remember the future and make plans for the past. Such a situation seems to our intuition quite nonsensical, but one important lesson of twentieth-century physics has been that we cannot always trust our intuition.

Even with this reservation, the proposition does not seem attractive. At any rate we can be sure that, if such beings existed, we would not be able to communicate with them, since communication necessarily involves the possibility of question and answer, and they would have to answer our questions before (from our point of view) they have been asked.

The discussion given so far applies to all statistical mechanics arising from experiments we can perform in the laboratory. But that is not the whole story. Around us, in the universe, many thermodynamic processes take place which obey the Second Law, but which are in no way influenced by our actions. There is sometimes some confusion about whether the Second Law applied to the universe, because one may ask the question in terms of the entropy of the universe as a whole. This is a concept inviting complications, because it is not certain whether the universe is finite, and also because there is some controversy about the way thermodynamic quantities behave under Lorentz transformations, so we may have some doubt about how to assess the entropy of rapidly moving objects.

But we can look at the question in a much simpler way, by considering only local behavior. In the sun, and even on the earth over regions too large to be affected by human intervention, the signs of all transport coefficients are the familiar ones. Heat is conducted from the hotter to the colder body, friction causes a force directed to make the velocity difference decrease in time, diffusion moves the components of a mixture so as to make the concentration gradient decrease with time, and so on. All these phenomena would be reversed if time went backwards.

Our previous explanation is obviously not adequate to explain this aspect of the arrow of time. Some authors have conjectured that the sense of the time direction has something to do with the expansion of the universe. This is not very plausible, because it is possible to visualize a situation in which the cosmological laws contain a minimum density, so that at some stage the expansion would stop and the universe would start contracting. This is probably not the true behavior of the universe, but it does not seem to contradict any fundamental laws of physics. It seems most unlikely that such a reversal would cause any change in the thermodynamic behavior of objects in the universe.

We need therefore a separate postulate to settle the cosmic arrow of time. Comparing the situation with the experiments discussed at the beginning, we see that the universe behaves as it would if it were the result of a gigantic experiment, started at some initial time with microscopic chaos, and allowed to run. This idea goes well with the "big-bang" hypothesis of the origin of the universe, but this should not be regarded as an argument for this version of cosmogony.

As before, we can ask whether it is an accident that our own arrow of time coincides with that of the universe, and whether it is conceivable that the two were opposite. This also seems quite unacceptable to our intuition. It is very strongly ingrained in our thought that the cause must come before the effect. This is part of the intuitive law of causality, which goes beyond the causality valid in mechanics or electrodynamics. The latter merely specifies that there exist differential equations for the time evolution of physical quantities, so that knowledge of a sufficient number of variables at one time determines their values for the future. It is not usually stressed that the same information determines their values in the past, so that this form of causality makes no distinction between past and future. Our intuition does, and while modern physics has in several instances caught out our intuition as misleading, this could conceivably be an instance where intuition leads us to a general truth which is not included in the basic laws of physics (except in the "derived" Second Law of Thermodynamics).

It should perhaps be pointed out that, in the interests of simplicity, I have chosen only purely mechanical examples, and not problems involving radiation. One could therefore ask how the usual procedure involving the use of retarded rather than advanced potentials relates to what we have discussed. Clearly the use of retarded potentials causes irreversibility, including radiative damping.

In selecting retarded potentials we make the assumption that at infinite distance there is no incoming radiation except for that specified as part of the problem, but there is no restriction on the outgoing radiation. This is equivalent with the conditions we have imposed on our mechanical problems, in assuming that we prepare a state of the system at $t = 0$ and observe the development after that. Indeed, if we start at $t = 0$ with no radiation, or with some prescribed radiation, we shall find that at later times some radiation will be traveling outward.

The situation is somewhat different if we adopt the ingenious description by Wheeler and Feynman (*Rev. Mod. Phys.*, 17, 157, 1945), who use a symmetric combination of retarded and advanced potentials, and rely for radiation damping on the effect of the advanced potentials from the atoms which will eventually absorb the radiation. In order to·obtain the same results as from the conventional theory, they have to assume that all radiation will ultimately reach an absorbing body. This could create difficulties if it turned out that there was not enough matter in the universe to make absorption certain. But quite apart from that problem, it is not often realized that this theory places the origin of the irreversibility in the absorbing matter. In the otherwise time-symmetric interplay of advanced and retarded potentials, the observed radiation damping requires that radiation normally encounter objects which obey the Second Law of thermodynamics, so that their behavior is irreversible.

The arguments discussed here were presented in a lecture by the author at a Birmingham symposium in 1967 (*Methods and Problems of Mathematical Physics*, ed. J. E. Bowcock, North-Holland, 1970, p. 3). Very similar views have been expressed by R. P. Feynman (*The Character of Physical Law*, BBC Publications, 1965, Lecture 5).

4. CONDENSED MATTER

4.1. MELTING IN ONE, TWO, AND THREE DIMENSIONS

One typical characteristic of a crystalline solid, as opposed to a liquid or an amorphous solid, is its periodic structure, by which the distance between two atoms in one of the principal crystallographic axes is close to a multiple of the basic spacing in that direction. One is always tempted to visualize such situations in the case of a one-dimensional system, because of its simplicity. Indeed, if we imagine a linear chain of atoms constrained to move in one dimension, and assume reasonable short-range forces between the atoms, the potential minimum, and hence the configuration at zero temperature, (neglecting quantum effects for the moment) will correspond to equal spacing. For example, if we assumed forces between nearest neighbors only, the equilibrium spacing a will be that for which the interaction potential between two neighbors is a minimum.

However, it is also clear that in this simple model there will be, at any finite temperature, deviations from the simple periodic structure which become cumulative with distance, so that the nth atom, which in equilibrium should be a distance na from that at the origin, will fluctuate around that position by an amount which increases with n, and for large n becomes greater than a. In the simple model we have postulated, this is very easy to see. The first neighbor of the reference atom will deviate from the equilibrium distance a by an amount δ, whose root mean square depends on the temperature and on the forces. In fact

$$\langle \delta^2 \rangle = KT \left(\frac{d^2 U}{dx^2} \right)^{-1}, \tag{4.1.1}$$

where U is the interaction potential and the derivative is to be taken at equilibrium.

Now the next atom is not influenced by the reference one, because we assumed only nearest-neighbor interaction; it will try to keep the correct distance from the first neighbor, again with an error of the same order. Evidently the position of the nth atom will have an error which is the superposition of n independent errors, each with the same distribution, so that the mean square of the error in the position of the nth atom will be $n\delta^2$, and the root mean square displacement grows as $n^{1/2}$, as in a random walk problem. When this quantity exceeds a, the periodic correlation in the positions will be lost.

The argument I have given here seems to depend on there being a force only between nearest neighbors, but we shall see presently that the conclusion is more general.

It was not at first realized that this behavior applies only to a one-dimensional system, and one is tempted to generalize it to real crystals. However, it is evident qualitatively that our reasoning does not apply in two or three dimensions. Take, for example, a square lattice in two dimensions. Even with nearest-neighbor forces only, the atom at (a, a) interacts with its neighbors at $(a, 0)$ and $(0, a)$, both of which interact with the atom at the origin. The atom at (a, a) will be encouraged to take up a wrong position only if both the intermediaries happen to be displaced in the same direction. With increasing distance from the origin there are more and more linkages by which information about the reference atom can reach the distant one. This tends to counteract the weakening of the correlations carried by each linkage. The number of such linkages grows even faster in three dimensions. While this reasoning shows that the answer in two and three dimensions is not necessarily the same as in one, it is hard to make it quantitative.

There is, however, a very simple way to find the solution, at least if the errors in the distance of more or less adjacent atoms are small enough to treat the forces as harmonic. This is a good approximation at low temperatures (except for very light and weakly bound atoms such as He and H_2) and reasonably correct even at the melting point.

In that case we can express the motion of the atoms in terms of normal coordinates. If u_n is the displacement of the nth atom from its equilibrium position, we can write, in one dimension:

$$u_n = \sum_k q_k \, e^{ikna}, \tag{4.1.2}$$

where q_k is the amplitude of the normal mode of wave number k, and the summation runs over all values of k which are multiples of $2\pi/L$, where L is the length of the chain, within the interval

$$-\frac{\pi}{a} < k \le \frac{\pi}{a}. \tag{4.1.3}$$

The reality of u_n requires that

$$q_{-k} = q_k^*. \tag{4.1.4}$$

We can now express the correlation between the positions of two distant atoms by considering the quantity $\langle (u_n - u_0)^2 \rangle$. If this is small it indicates that the distance between the two atoms is close to na. If the quantity becomes larger than a^2, the coherence in position is lost.

We can easily evaluate this mean square fluctuation from (4.1.2) by using the fact that, from the equipartition law,

$$\langle q_k q_{-k'} \rangle = \begin{cases} KT/M\omega_k^2 & k' = k, \\ 0 & k' \ne k, \end{cases} \tag{4.1.5}$$

where K is again Boltzmann's constant, and M the mass of the whole chain. ω_k is the frequency of the mode k. For a long chain the k values are very dense, so the sum may be replaced by an integration, with the weight factor $(L/2\pi)dk$. Finally we have

$$\langle (u_n - u_0)^2 \rangle = \frac{KT}{2\pi m} \int_{-\pi/a}^{\pi/a} dk \, \frac{1 - \cos kna}{\omega_k^2} \tag{4.1.6}$$

Here m is the mass of one atom.

In the region of large k the cos term in the integrand oscillates and does not contribute much to the integration. However, for small k, where ω_k is approximately ck, c being the velocity of sound, the integral would diverge without the cos term. With the cos term, the integrand tends to the finite value $n^2 a^2/2c^2$ at the origin, and since it remains of this magnitude over an interval of the order of $1/na$, we see that the mean square fluctuation becomes, for large n, proportional to n. This agrees with our previous, more elementary, argument, but the result is now seen to be independent of the assumption that only nearest neighbors interact, provided only the interaction falls off sufficiently rapidly with distance to give a finite velocity of sound for long waves.

In the physical problem of a three-dimensional crystal, one obtains an expression very similar to (4.1.6), complicated a little by our having to pay attention to the polarization of each mode, and to the direction of the wave vector, of which only the component along the line joining the two atoms enters in the argument of the cos. However, the main difference is that the **k** integration is now in three dimensions, with a volume element which is proportional to $k^2 \, dk$. This compensates the small ω_k^2 in the denominator, and therefore the two terms in the integrand contribute finite amounts to the integral. As a result, the mean square fluctuation tends to a finite limit for large n; this limit is proportional to T, so for low enough temperature the quantity must be less than a^2.

I believe this argument for the existence of long-range order in three dimensions was first given in a paper of mine, *Helv. Phys. Acta*, 7, Suppl. 2, p. 81, 1934. (See also *Ann. Inst. H. Poincaré*, 5, 177, 1935.) Once it is established that there is long-range order at low temperature, it follows that the thermodynamic quantities cannot be analytic functions of T over the whole range. The point is that we know, for example from series expansions in inverse powers of T, that at high enough temperatures there is no long-range order. But a function which is zero over a range of temperatures, and non-zero at the lower temperatures, cannot be wholly analytic; there has to be a lowest temperature at which the order is zero, and this requires a singularity. This very general argument does not tell us whether the order is discontinuous, suggesting a first-order transition (as is the case for real solids), or whether only its first, or even a higher, derivative is discontinuous. There are more powerful arguments due to Landau based on the disappearance of translational symmetry on solidification, which settle this point.

The two-dimensional case, in which the element of integration is $k\,dk$, leads to an expression which increases as $\log n$. This led to the conclusion that there would, in two dimensions, be no long-range order, and therefore, presumably, no sharp melting point. The argument was later refined and generalized by D. Mermin (*Phys. Rev.*, 176, 250, 1968).

It therefore came as a surprise when experiments on adsorption of helium on various surfaces, using less than a monolayer, showed a behavior strongly suggesting a solid and a liquid phase with a fairly sharp transition. There was no direct proof that one was indeed dealing with a solid-liquid transition, and the periodic structure of the substrate and the oscillations of the helium atoms perpendicularly to the surface made the system somewhat different from the interacting atoms moving in a plane to which the theory applied. Nevertheless, it was an occasion to re-examine the old argument against a melting point in two dimensions.

This led to a consideration of orders of magnitude. Since the mean square fluctuation grows with n, it will reach the value a^2 for some finite n. But since the increase is only logarithmic, with a temperature-dependent factor outside the logarithm, it is easily seen that the correlation length, or the distance over which there is periodicity in the atomic positions, contains an exponential with an exponent which can be very large at low temperatures. The distance over which order is maintained, while finite, can become extremely large, and if it exceeds the size of the specimens used in practice, the fact that it is finite should cease to be of importance.

In any finite system the transition is not sharp, but spread over a temperature interval which depends on the size of the system. One conjectures that in the idealized case of an infinite two-dimensional system the transition would be spread over a temperature interval characteristic of the correlation length. The transitions seen in the experiments were broad enough to be consistent with all this, and the width of the transition was probably due to inhomogeneities and other experimental factors.

At this stage it seemed that the theoretical conclusion that there was no long-range order and no sharp melting point in two dimensions was correct, but practically irrelevant. But there was yet another surprise to come, in the paper by Kosterlitz and Thouless, *J. Phys.*, C6, 1181, 1973. For our present purpose, their argument can be re-stated as follows.

The symmetry of a crystal manifests itself normally in two related, but separate, ways. One is the periodicity in the positions of the atoms, which is shown by sharp X-rays or neutron diffraction lines. The other is the persistence of crystallographic directions, so that different faces of a crystal tend to be parallel or to form discrete angles with each other. (In practice, this latter property is what we notice, and admire, in crystals.)

One aspect of long-range order in a perfect crystal is therefore coherence in direction. Considering a particular atom and one of its nearest neighbors, we can define a reference axis by the line joining them. If we then take another atom n spacings distant, we can look for the lines joining it to its neighbors and find the one closest in direction to the reference axis. The angle θ between this and the reference axis would be zero for perfect order. If $\langle \theta^2 \rangle$ is smaller than its random value, even in the limit of large n, we have directional long-range order. Our discussion, which started from the one-dimensional chain as its pattern, did not refer to this, because in one dimension there are no directions. We therefore did not examine the question of directional order in two dimensions, and the argument sketched above does not exclude this.

In fact, it is possible to write an equation resembling (4.1.6) for the mean square of θ, and this tends, at infinite n, to a finite limit which decreases at low temperature. Kosterlitz and Thouless give a more general argument: If there is high local order, i.e., if the distances of each atom from its neighbors do not differ much from those in the perfect lattice, then it is impossible to make the crystallographic axes change by an appreciable angle, even over large distances. The only way this can be achieved is by including dislocations, i.e., places in which two rows of atoms get out of step by one containing one more atom than the other. This involves a configuration in which the local order has been strongly disturbed.

The energy of a dislocation is finite, and at low temperature the density of dislocations is therefore exponentially small. However, the energy required to form further dislocations decreases as the dislocation density increases. As the temperature increases the dislocation density will reach a value at which further dislocations form easily, and at this point the directional order disappears. Apart from re-establishing the existence of a sharp transition in two dimensions, the theory has therefore given us a new picture of the mechanism of melting, which probably is applicable also in three dimensions.

4.2.   MOMENTUM OF PHONONS

It is well known that the thermal motion of the atoms of a crystal lattice consists of lattice vibrations which, to a reasonable approximation, can be regarded as harmonic. The quanta of this motion are the phonons, in the same way in which the quanta of the electromagnetic field are called photons.

One sometimes finds the statement that a phonon of wave vector $\mathbf{k}$ has momentum $\hbar\mathbf{k}$. Since its energy is $\hbar\omega$, or $\hbar c_\phi k$, where c_ϕ is the phase velocity, the statement amounts to saying that the ratio of momentum to energy is the reciprocal of the phase velocity, and in that form has meaning also in the classical case. Is it a correct statement?

As long as the amplitude of the vibrations is small enough to treat the equations of motion as linear, the most convenient variables of the problem are the normal coordinates q_k, which we have already used in the one-dimensional problem in (4.1.2). In the harmonic case these are arbitrary and independent of each other. In terms of these the momentum is

$$p = \sum_n m\dot{u}_n = \sum_k m\dot{q}_k \, e^{ikna}, \qquad (4.2.1)$$

where m is the atomic mass. But the last sum vanishes except when $k = 0$. This follows because the allowed values of k are such that kx runs through a whole number of cycles from one end of the chain to another. At least this is the position when we use a cyclic boundary condition. For a finite chain the normal modes become standing waves, but they have to be orthogonal to the $k = 0$ mode, which represents a uniform displacement of the whole chain and thus is a possible normal mode of frequency zero. Hence the whole momentum comes from this one mode, which is not a proper phonon mode since it has no restoring force, and the proper phonon modes do not contribute to the momentum. The same is true in three dimensions and for more complicated lattices.

The reason why one is inclined to try to attribute a momentum $\hbar\mathbf{k}$ to a phonon is that there is a kind of conservation law for the sums of the wave vectors of the phonons and those of electrons, photons, and other objects with which they interact. This is because the quantity $\hbar\mathbf{k}$, while not a momentum, is what is often called pseudomomentum, or crystal momentum.

Its physical significance is related to the translational symmetry of the crystal. Every conservation law in physics is connected with a symmetry; momentum conservation depends on the homogeneity of space, i.e., on the invariance of the equations under the substitution

$$x,y,z \rightarrow x + b,y,z \qquad (4.2.2)$$

for any b. If this invariance does not hold, for example because there is a fixed center of force, momentum is not conserved.

Pseudomomentum conservation reflects a different symmetry,

namely, the invariance of the equations under a change in which all physical attributes of any one element of the medium (mechanical displacement, density, electric and magnetic fields, etc.) are transferred to another element a distance b away:

$$f(x,y,z) \to f(x - b,y,z) \qquad (4.2.3)$$

for any quantity f. The condition for this invariance is uniformity of the medium. The difference between (4.2.2) and (4.2.3) is that in the latter the medium is not moved. In the absence of any medium, for example for free particles, or for light in vacuum, there is no difference between momentum and pseudomomentum.

In the case of a crystal, (4.2.3) does leave the equations unchanged if we can disregard the boundaries, provided b is a multiple of the lattice spacing. The fact that the displacement cannot be made arbitrarily small has the consequence that pseudomomentum is not exactly conserved, but can change by amounts of the form $\hbar\mathbf{G}$, where $\mathbf{G}$ is an arbitrary vector in the reciprocal lattice of the crystal. For example, in the linear chain we have discussed, the definition (4.1.2) of the normal coordinates is not changed if any multiple of $2\pi/a$ is added to any of the wave vectors; multiples of $2\pi/a$ form the reciprocal lattice of this one-dimensional lattice.

In many problems of interest, both momentum and pseudomomentum are conserved. Take as a typical example the creation of a phonon by the scattering of a neutron by a crystal. If this takes place far enough from a boundary, the crystal is uniform, and pseudomomentum is conserved. Let the initial and final momentum of the neutron be $\mathbf{p}$ and $\mathbf{p}'$. Since the neutron interacts only weakly with matter, it is essentially a free particle, and its momentum and pseudomomentum are equal. The phonon which is produced, therefore, has pseudomomentum $\hbar\mathbf{k}$, where

$$\hbar\mathbf{k} = \mathbf{p} - \mathbf{p}'. \qquad (4.2.4)$$

This simple selection rule makes neutrons such good tools for studying phonons. But the total momentum is also conserved, and therefore the crystal (or if it is supported, the support) must receive the momentum lost by the neutron, equal to (4.2.4). From what was said above, this means that in addition to the q_k mode, the q_0 mode must also acquire a certain amount of excitation.

If, together with the creation of a phonon, the crystal always receives momentum $\hbar\mathbf{k}$, is it perhaps a matter of definition whether we count this momentum as belonging to the phonon? If this were reasonable, we would expect that if we knew the position of the phonon (to be sure, approximately, within the limits set by the uncertainty principle) the momentum should be localized in the same place. This, however, will not in general be possible, because of the phenomenon of dispersion, i.e., the dependence of the velocity of propagation on wavelength.

If the cause of phonon emission, e.g., the neutron, is initially localized, both the oscillatory motion of the lattice and the unidirectional motion carrying the forward momentum will be localized in the same region, giving a velocity distribution of the form of Figure 4.1a. However, after some little time, the oscillatory part, i.e., the phonon proper, will have progressed with its group velocity, whereas the forward motion, which consists of much longer waves, will have moved ahead with the velocity of sound, which is greater. The velocity distribution will, after some time, look like Figure

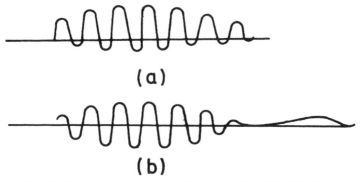

(a)

(b)

Figure 4.1   Lattice vibration caused by a localized periodic force: (a) initially, (b) after some time.

4.1b, and this makes it clearly unreasonable to regard the momentum-carrying part as part of the phonon.

Some care is necessary in using this argument, because a wave packet of finite size necessarily spreads as time goes on, and it has to be shown that by the time the two distributions have separated by more than their diameter, as shown in the figure, the spreading has not grown so large as to bridge the gap. A dimensional consideration shows, however, that this condition is satisfied provided the phonon wavelength is not very large compared to the lattice spacing, and the initial extension of the wave packet is much larger. The figure is strictly applicable only to a classical situation, but the answer translates exactly into quantum mechanics.

We therefore conclude that, at least in the harmonic approximation, a definition which attributed the momentum to the phonon would be physically unreasonable.

In many interaction processes between solids and radiation, such as the Brillouin scattering or the Cherenkov radiation, the arguments usually given in terms of momentum are correct if interpreted in terms of pseudomomentum conservation. This is just as well, because the actual momentum of light in a refractive medium is a very complicated quantity. It is fortunate that we are not usually interested in the total momentum of the solid.

While the difference between momentum and pseudomomentum is familiar to most physicists concerned with solids, it is customary in the theory of liquids to assume that phonons carry momentum. It is therefore interesting to consider the problem of a sound wave in a liquid. We know that there are two different, but equivalent, ways of describing the motion of a liquid: that of Euler, and that of Lagrange.

In the Lagrange variables, one states the position at time t of the element of fluid which, at some reference time, say $t = 0$, was at some point $\mathbf{R}$. So one is dealing with a vector function

$$\mathbf{r}(\mathbf{R}, t). \tag{4.2.5}$$

This is just the continuum limit of the natural variables for the description of lattice vibrations. The displacements $\mathbf{r} - \mathbf{R}$ are the continuum limit of the u_n occurring, for example, in (4.1.2). These variables are rather inconvenient for the general motion of a liquid because there is no restoring force for transverse displacements, and the displacements can easily get very large, so that the problem of discovering where a particular fluid element has got to may become quite complicated. For the case of a longitudinal sound wave in an otherwise stationary fluid, the equations are quite simple, and, if linearized for small amplitude, are identical with the equations for an elastic wave in a solid. Not surprisingly, therefore, we still find zero momentum.

We can formulate the same problem in Euler variables, in which one states the velocity $\mathbf{u}$ and the density ρ as a function of the space coordinates and time. The equations of motion of a non-viscous fluid are then

$$\dot{\mathbf{u}} + (\mathbf{u} \cdot \mathbf{V}) + \frac{1}{\rho} \mathbf{V}p = 0,$$

$$\dot{\rho} + \mathbf{V} \cdot (\rho \mathbf{u}) = 0. \qquad (4.2.6)$$

For a sound wave of small amplitude $\mathbf{u}$ is small, and the deviation from normal density is also small. If we write

$$\rho = \rho_0(1 + \eta), \qquad (4.2.7)$$

we may treat $\mathbf{u}$ and η as small of first order and neglect second-order terms. This leaves

$$\dot{\mathbf{u}} + c^2 \mathbf{V}\eta = 0,$$

$$\dot{\eta} + \mathbf{V} \cdot \mathbf{u} = 0, \qquad (4.2.8)$$

with the abbreviation

$$c^2 = \frac{dp}{d\rho}, \qquad (4.2.9)$$

where c is the velocity of sound. A typical solution of (4.2.8) is

$$u_x = A \cos k(x - ct), \quad \eta = \frac{A}{c} \cos k(x - ct). \qquad (4.2.10)$$

The momentum density is

$$\rho u_x = \rho_0(1 + \eta)u_x,$$

and its space or time average is $\rho_0 A^2/2c$, which is $1/c$ times the energy density. (It is evident that the density of kinetic energy is, to the leading order, on the average, $\frac{1}{4}\rho_0 A^2$, and one knows that for harmonic motion the averages of kinetic and potential energy are equal.)

If this motion is quantized, the energy per phonon is $\hbar\omega = \hbar ck$, and if the momentum-to-energy ratio is $1/c$, this gives each phonon a momentum of $\hbar\mathbf{k}$. Thus we find the surprising result that a phonon in a liquid should have no momentum, or momentum $\hbar\mathbf{k}$, according to our choice between two equivalent sets of variables.

To understand this paradox one has to remember that the transformation from Euler to Lagrange variables is not linear. For small amplitude the first-order terms are equivalent in both descriptions, but if in the Euler variables we choose the solution (4.2.10), which contains only a wave of wave vector $\mathbf{k}$, and no uniform flow, this will *not* transform in Lagrange variables into a state in which only the kth mode is excited, but there is also a velocity contribution, small of second order, from the $\mathbf{q}_0$ mode. Conversely, if we start with motion given by only a single mode in Lagrange language, the translation will not be (4.2.10), but will contain in addition a small backwards flow, which cancels out the momentum we have calculated.

This situation is possible only because the description of a sound pulse in a liquid is not unique. In the equations we have used, there is no dispersion, i.e., the velocity of propagation is independent of wavelength; it is the same for the short-wave sound wave as for a slowly varying velocity field. If we add a uniform background velocity to the sound pulse, as in Figure 4.1a, it will travel with the sound pulse and will not separate as in Figure 4.1b. It is then a matter of definition how we choose to picture the sound wave, and the Euler and Lagrange variables merely suggest different choices as being simple.

We should therefore not be surprised to find that in the case of continuous, and therefore dispersion-free, liquids both points of view are consistent. In the two-fluid theory of liquid helium, in particular, it is customary to define phonons as carrying the momentum $\hbar\mathbf{k}$. In the usual presentation one gains the impression that this is essential for the theory, which would be surprising, in view of our conclusion that it is a matter of definition. This question has been studied by Thellung, who found that the equations of the two-fluid model could be re-written by changing variables in such a way that phonons carry no momentum. In this re-formulation all physical results remain unchanged, but the definition of the division of the flow into a normal and a superfluid part has to be adjusted. This argument by Thellung is unpublished, but it is referred to in a foot-note in his paper in *Supplemento al Nuovo Cimento*, 9, 243, 1958.

In real fluids there is, of course, dispersion. Even in liquid helium a variation of velocity with wave vector has been found. We must then expect that the freedom of choice is lost, and that we are no longer free to define the state of a phonon with an arbitrary momentum. If we are guided by the analogy with phonons in crystals, we are tempted to believe that the momentum should again be taken to be zero. This would be the case if we could be sure that the Lagrange equations of motion are still linear, which, as we have seen, implies that the Euler equations are not.

Now we know that both in crystals and in liquids the harmonic theory is only an approximation, and that in reality there are non-linear terms in the equations of motion. These have been studied for crystals, and are responsible, for example, for the finite thermal conductivity of perfect crystals. It is known that they have the consequences of limiting the lifetime of a free phonon, by processes in which this phonon splits in two or more, or combines with one or more other phonons, if more are present. This may cause a limitation to our argument if the lifetime of a phonon becomes shorter than the time in which the wave packets of Figure 4.1 separate cleanly. Estimates show, however, that, at least at low temperatures, this is not a serious limitation. To the extent to which this effect of anharmonicities has to be allowed for, it essentially limits the accuracy with which a one-phonon state can be defined.

It is conceivable that non-linearities can also have a different effect. To see this, note that in the academic, but simple case of a liquid for which the Lagrange equations are exactly linear but have dispersion, the Euler equations are, as we have seen, non-linear. In Euler language we are therefore dealing with non-linear equations, which would not, in general, permit a phonon-type solution in which there is a strictly periodic motion. A function of the type (4.2.10) is then only an approximate solution. However, we know, from the findings in the Lagrange language, that there exists a solution which is strictly periodic and carries no momentum. In the Euler language this must contain, apart from the motion (4.2.10), additional small contributions of long wavelength which will cancel out the momentum, and lead to a strictly periodic solution of the non-linear equations. Is it possible that non-linearities in the Lagrange equations could have a similar effect? We can hardly expect a strictly periodic solution in the presence of the general anharmonicities, but it could be that a more nearly periodic solution, representing a phonon of longer life, would require corrections which might have contributions from the q_0 mode, and thus a non-zero momentum. I do not know the answer to this question, so that further surprises are not ruled out.

4.3. Electron Diamagnetism

The question of the behavior of free electrons in an external magnetic field has long been of interest. Since the electron orbits are helical, their projections on a plane at right angles to the field being circles, one expects that the electrons might have non-vanishing magnetic moments, even without allowing for spin.

A naive approach would say that, since the radius of the projected orbit is $v/2\omega_L$, where v is the velocity in the plane perpendicular to the field, and ω_L the Larmor frequency, so that

$$r = \frac{mv}{eB},$$ (4.3.1)

the magnetic moment of the orbit should be evr, or

$$\mu = \frac{mv^2}{B}.$$ (4.3.2)

This answer is clearly nonsense. It does not depend on the electron charge and decreases with increasing field, so that it would become infinite for $B = 0$!

The mistake comes from the fact that we are talking in terms of complete orbits. If we are dealing with a finite region of space, we should not consider the orbits whose centers are within that region, but the electrons which at a given time actually are in the region, wherever the center of their orbits may be. For $B \to 0$, when (4.3.2) goes wrong so badly, the orbits are practically straight, and the centers very far away. The error was explained, and corrected long ago, by H. A. Lorentz, quoted by Miss van Leuven (*J. de Physique*, 2, 361, 1921), and also by Niels Bohr in his 1911 Copenhagen thesis.

Figure 4.2 shows a rectangular area with a number of circular orbits. Some are entirely within the area. Some are entirely outside, and we are not interested in them. Some intersect the boundary, and a sequence of them has been shown. We may regard the rectangle just as a reference area for counting electrons; in that case the electrons go in and out, but we want to include them only when they are inside, i.e., when they are on the sections of the orbits shown in a heavy line. Or we may regard the rectangle as a box with reflecting walls, in which case the heavy lines form one orbit of an electron which keeps being reflected from the wall. The net effect on the current distribution is the same, and it is more convenient to talk in terms of the second alternative. In that case the heavy lines form one large orbit, which is gone through in an anti-clockwise direc-

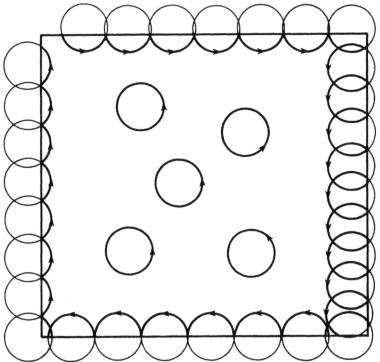

Figure 4.2   Two-dimensional motion of electrons in a rectangular box in a magnetic field. The internal orbits have a small magnetic moment in a clockwise sense, the boundary orbit shown has a large moment in an anti-clockwise sense.

tion if the motion along the circular orbits is clockwise. If the electrons are distributed at random, the number of orbits intersecting the boundary is relatively small, but the area enclosed, and therefore the magnetic moment, is large, and it is therefore plausible that they can cancel the magnetic moment of the circular orbits which do not touch the wall. It is possible to verify this, but the calculation is not very convenient.

There is a simpler, and much more general, argument based on statistical mechanics. We know that the magnetic moment of a system in statistical equilibrium is given by

$$\mu = -\frac{\partial F}{\partial B},\qquad (4.3.3)$$

where F is the Helmholtz free energy. In Boltzmann statistics, the free energy per electron is given by

$$Z = e^{-\beta F} = \int e^{-\beta H(\mathbf{r}, \mathbf{p})} \, d^3\mathbf{r} \, d^3\mathbf{p}, \qquad (4.3.4)$$

where, as usual, β is $1/KT$, and H is the Hamiltonian. The integral is to be taken over all phase space. In the magnetic field,

$$H = V(\mathbf{r}) + \frac{1}{2m}(\mathbf{p} - e\mathbf{A})^2, \qquad (4.3.5)$$

where V is the potential energy, including the repulsion due to the wall, if any, and $\mathbf{A}$ the vector potential. If we insert (4.3.5) in (4.3.4) and change the integration variable from $\mathbf{p}$ to

$$\mathbf{q} = \mathbf{p} - e\mathbf{A}, \qquad (4.3.6)$$

the integral is seen to be independent of $\mathbf{A}$, and therefore the magnetization (4.3.3) vanishes.

The reason why this way of calculating the magnetic moment is better than the first one is that we carry out the integration for the partition function (4.3.4) *before* differentiating with respect to B. Z is not very sensitive to the boundary, except that the integration stops there. If we differentiate inside the integral, and therefore follow the change of each orbit with the magnetic field, then orbits close to the boundary change their nature completely as, with a change of radius, they intersect the boundary. This causes the sensitivity expressed in the figure.

Our conclusion that in classical mechanics there can be no magnetization is also valid if there are interacting particles, and it extends to the case of Fermi or Bose statistics. It does not, of course, apply to considerations such as Curie's, in which one postulates that atoms have intrinsic magnetic moments, without attempting to explain their origin.

The very well-known argument sketched above is essentially classical, and one expected that the answer would be different in quantum mechanics. There was initially some hesitation in applying quantum mechanics to this problem, because the failure of the "naive" argument, and the situation illustrated in the figure, seemed to suggest that it was essential to treat the boundaries correctly. The solution of the Schrödinger equation in a box in the presence of a magnetic field is quite complicated.

It therefore came as a pleasant surprise when Landau showed that the answer could be obtained quite easily without worrying about the boundary effects in detail. One should not have been surprised—and Landau was not surprised, because he understood the advantage of computing the free energy and then using the identity (4.3.3). However, many people expressed doubts about the insensitivity of this procedure, and it was some time before Landau's result was generally accepted.

To obtain Landau's answer, we first consider the motion in a plane at right angles to the magnetic field. Because the classical motion is simply periodic, and the equations of motion linear, one can deduce immediately that the energy eigenvalues are

$$E_n = (n + \tfrac{1}{2})\frac{\hbar eB}{m} = (2n + 1)\hbar\omega_L = (2n + 1)\mu_B B, \quad (4.3.7)$$

$$n = 0, 1, 2, \ldots,$$

where ω_L is the Larmor frequency and μ_B the Bohr magneton.

In an infinite plane there is an infinite degeneracy, i.e., there are infinitely many states belonging to each of the energy eigenvalues (4.3.7), corresponding to the fact that the classical circular orbit may be located anywhere in the plane. If the area is limited, many of the states will be far from the boundaries, and their energies will still be given by the same formula. Those located outside do not exist (or have a very high energy because of the repulsive potential of the walls). A small fraction, located close to the wall, will have their energy shifted, but they are statistically insignificant. This is the point at which the device of computing the partition function before differentiating with respect to B pays off, and it was on this point that the early skeptics of Landau's result concentrated.

The degree of degeneracy can be deduced from the requirement that the number of states in an energy interval large compared to the spacing must still be the same as in the absence of the magnetic field, and this leads to the conclusion that there are, disregarding spin,

$$G = \frac{eBA}{2\pi\hbar} \tag{4.3.8}$$

states with the same energy E_n in a region of area A. This result can also be verified by a consideration of the two-dimensional Schrödinger equation in a magnetic field. (See Landau's original paper, or Peierls, *Quantum Theory of Solids*, 1955, §7.2).

Adding the kinetic energy of motion in the direction of the magnetic field, we find for the free energy, in the case of Boltzmann statistics:

$$F = -KT \log Z \, ;$$

$$
\begin{aligned}
Z &= \frac{LG}{2\pi\hbar} \int dp \, e^{-\beta p^2/2m} \sum_{n=0}^{\infty} e^{-(2n+1)\beta\mu_B B} \\
&= \frac{ALeB}{4\pi\hbar} \sqrt{\left(\frac{2\pi m}{\beta}\right)} \frac{1}{\sinh \beta\mu_B B}.
\end{aligned}
\tag{4.3.9}
$$

Here L is the length of the enclosure in the field direction. In practice, the value of $\beta\mu_B B$ is usually much less than unity, and we may expand the sinh. The leading term is easily seen to give the partition function for one free electron in a volume AL. The next term contributes a factor

$$1 - \tfrac{1}{6}(\beta\mu_B B)^2$$

to Z, and therefore a term $(\mu_B B)^2/6KT$ to the free energy, hence by (4.3.3) a magnetization of

$$-\frac{\mu_B{}^2 B}{3KT} \qquad\qquad (4.3.10)$$

per electron. This is just one-third of the paramagnetism due to the electron spin.

In metals the electrons usually form a degenerate Fermi gas, and one therefore has to use the expression for Fermi statistics in place of (4.3.4). The evaluation is straightforward, and it is again true that the diamagnetic susceptibility due to the electron orbits is one-third of the Pauli paramagnetism of a degenerate electron gas. This ratio is therefore independent of whether the electron gas is degenerate or not. If the diamagnetism is less than the paramagnetism, one might expect that metals could never, in the net result, be diamagnetic.

However, if the electrons are not free, but move in the periodic potential field of a lattice, their effective mass may be different from that of a free electron, so that the level spacing (4.3.7) will be changed. This is because it is the effective mass which determines the orbital motion in the field, whereas the spin magnetic moment remains the same as that of a free electron. One can understand why, for example in Bi, the energy shows, as a function of wave vector, a very large curvature, and therefore a small effective mass. This accounts for the large diamagnetic susceptibility of Bi.

The surprise lies, in this case, in the ease with which the problem can be discussed, in spite of its apparent complexity and sensitivity.

4.4.   DE HAAS-VAN ALPHEN EFFECT

The discussion of the previous item relied on expanding the partition function in powers of the small quantity $\mu_B B/KT$. If this quantity becomes somewhat larger, one may have to include terms of higher order, which will change the results quantitatively, but do not bring in any new features. But the use of such an expansion fails to bring out a very interesting effect, which becomes appreciable when $\mu_B B/KT$ is not too small.

To understand the origin of this effect without much calculation, let us return for the moment to the two-dimensional case, and also let us consider the limiting case of zero temperature, when energy and free energy are identical. Let us start with the situation in which the magnetic field is so strong that the degree of degeneracy (4.3.8) is greater than the number N of electrons present. Then, even with Fermi statistics, all electrons are in the lowest state $n = 0$, and we have

$$E = N\mu_B B, \quad \mu = -N\mu_B, \quad \text{if} \quad B > b, \tag{4.4.1}$$

where

$$b = \frac{2\pi\hbar N}{eA}. \tag{4.4.2}$$

If the field drops below this limit, $N - G$ electrons have to go into orbits with $n = 1$. This makes the energy

$$E = \mu_B B G + 3\mu_B B(N - G)$$
$$= \left(3N - \frac{2NB}{b}\right)\mu_B B, \quad \frac{b}{2} < B < b. \tag{4.4.3}$$

Then the magnetization becomes

$$-\frac{dE}{dB} = -\left(3 - 4\frac{B}{b}\right)N\mu_B, \quad \tfrac{1}{2}b < B < b. \tag{4.4.4}$$

As B varies through the stated interval, the magnetic moment varies from $-N\mu_B$ to $+N\mu_B$. A little algebra shows that when

$$\frac{1}{n+1}b < B < \frac{1}{n}b, \tag{4.4.5}$$

the magnetic moment is

$$\left[2n(n + 1)\frac{B}{b} - (2n + 1) \right]N\mu_B. \qquad (4.4.6)$$

Between the limits of the interval (4.4.5) this again changes from $-N\mu_B$ to $+N\mu_B$. We find violent discontinuous oscillations, the discontinuities being equidistant in the variable $1/B$.

The simple calculation reproduced above is, of course, very unrealistic, and for any practical conclusions one has to allow for a finite temperature and for the third dimension. Both these changes have the effect of smoothing the oscillations, but it remains true that there are oscillations, and the maxima or minima occur at equidistant values of $1/B$.

For a full calculation one has to evaluate the expression for the free energy of a Fermi gas

$$F = N\eta - 2KT \int_{-\infty}^{\infty} dE \log\left[1 + e^{-\beta(E - \eta)} \right]g(E), \qquad (4.4.7)$$

where η is the Fermi energy, and $g(E)$ the density of states, in our case

$$g(E) = \frac{2(2m)^{1/2}VeB}{(2\pi\hbar)^3} \sum_n \left[E - (2n + 1)\mu_B B \right]^{1/2} \qquad (4.4.8)$$

where $V = AL$ is the volume, and the sum over n is to be taken over all those values for which the square root is real.

If we replaced the sum in (4.4.8) by an integral we would get back to the classical result, with no diamagnetism. To approximate the difference between sum and integral it is convenient to make use of Poisson's summation formula, which is not as widely known as it deserves to be. For any sum of the form

$$S = \sum_{n-\infty}^{\infty} \phi(n)$$

we can write

$$S = \int_{-\infty}^{\infty} \phi(x)D(x)\,dx, \quad D(x) = \sum_{n=-\infty}^{\infty} \delta(x-n). \qquad (4.4.9)$$

$D(x)$ is a periodic function of period 1, and can therefore be expanded in a Fourier series:

$$D(x) = \sum_{\ell=-\infty}^{\infty} e^{2\pi i \ell x}. \qquad (4.4.10)$$

We can now insert this expression in the integral (4.4.9) for S and interchange the order of summation and integration. (Purists will be shocked by these proceedings since both the Fourier representation of the singular function $D(x)$ and the change in the order of the operations are of doubtful validity. As happens so often, however, in dealing with Fourier transforms and with the Dirac δ-function, the result obtained after *both* these dubious operations can be justified in a wide range of circumstances.)

If this expansion is inserted in (4.4.7) we obtain, apart from terms independent of the magnetic field, and the term giving the steady diamagnetism already discussed, a contribution to the free energy which contains the de Haas-van Alphen effect:

$$F_{HA} = 2\pi V \frac{(2m)^{3/2}}{(2\pi\hbar)^3} \sum_{\ell=1}^{\infty} \frac{\cos\left[\dfrac{\pi\ell\eta}{\mu_B B} - \dfrac{\pi}{4}\right]}{\sinh(\pi^2 \ell KT/\mu_B B)}. \qquad (4.4.11)$$

For any reasonable field strength and reasonable temperature the denominator increases very rapidly with ℓ, so that usually only the first term matters. This has in the numerator an oscillating function of B^{-1}, as expected, and its amplitude, determined by the denominator, decreases exponentially as the ratio $\mu_B B/KT$ becomes small.

The surprising feature of this situation is that the conventional reaction of the theoretician to a complicated problem containing a small parameter, namely, to expand in powers of this parameter, fails completely. The function

$$F(z) = e^{-1/z}$$

has the property that, for real positive z, all its derivatives tend to zero as $z \to 0$. The attempt to construct a Taylor series for $F(z)$ therefore leads to a series in which every term vanishes identically. There is no question of the convergence of this series; in fact it is the most rapidly convergent series there is, but it has nothing to do with the function $F(z)$. In this case, when we are given the explicit form of F, we see immediately that it has an essential singularity at $z = 0$, but if F turns up in an intricate physical problem this may not be evident.

If a function of this form is the whole answer, the fact that its Taylor series is identically zero would probably suggest that there is something singular going on. However, frequently, as in our present problem, there are also other contributions, which can be expanded, and therefore there is nothing strange about the appearance of the series.

One is saved from the error by starting from the simple discussion of the extreme case of very strong field and zero temperature, which was our starting point, and which leads one to expect an oscillatory behavior. One then looks for a method of approximation in which these oscillatory terms would not get lost.

There is more than academic interest in this problem because in real metals the oscillations reveal a great deal of information about the structure of the Fermi surface in wave vector space, but this is not the place to discuss the methods or results of this approach.

Historical note: The oscillatory behavior was noticed by Landau in his first paper on diamagnetism, but he regarded it as unobservable in practice. The discovery of the oscillations in Bi by de Haas and van Alphen therefore seemed a complete mystery. The present author, having missed or forgotten Landau's remark, then suggested the quantized orbits as the origin of the effect, and illustrated this by some rough numerical calculations, which were later extended by Blackman. The use of the Poisson summation formula was suggested by Landau. For a fuller account see D. Shoenberg (to be published).

5. TRANSPORT PROBLEMS

5.1. DENSITY EXPANSION OF DIFFUSION COEFFICIENT

The next example concerns a very basic point in the statistical mechanics of gases. It is relevant both to classical and to quantum theory, but for simplicity I shall discuss only the classical case.

It is well known how simple kinetic theory handles such non-equilibrium problems as the viscosity of a nearly perfect gas. Following Boltzmann we consider binary collisions, in which molecules coming from different places, and therefore carrying, on the average, velocities appropriate to different points in the velocity field, tend to equalize their velocities, and thereby to diminish the velocity gradient. The simple treatment applies to very dilute gases, in which the restriction to binary collisions is justified because the chance of three or more atoms being within the range of each other's forces is proportional to a higher power of the density, and is therefore by definition negligible.

There might seem to be some danger in this statement because one is usually interested in a macroscopic problem, in which the total number of molecules is very large. In such a large system the probability that some molecules are in interaction with two or more others is, in fact, high. What actually matters is that any one molecule has a small chance of being simultaneously in interaction with more than one other.

Some care is therefore necessary to write the theory in such a form that it is seen to depend only on the latter approximation, which is justified, and not on the assumption that the total number of ternary and higher collisions in the volume is small. For equilibrium problems this is achieved by the "linked-cluster" formalism, to which we shall return later. For the viscosity a similar expansion was developed by Bogolyubov in 1946 and this remained the standard reference on the subject for many years. Evidently the many people who quoted Bogolyubov's expansion had never looked in detail at more than the first two terms of this expansion.

It was then one of the major surprises in theoretical physics when the work of Dorfman and Cohen (*Physics Letters*, 16, 124, 1965) showed that this expansion did not exist. The point is not that it diverges, the usual hazard of series expansions, but that its individual terms, beyond a certain order, are infinite.

To understand the nature of this surprise, we shall consider, once again, the so-called Lorentz gas. This is a model in which the molecules do not collide with each other, but with fixed scattering center distributed at random. This simplifies the kinematics greatly, without changing the nature of the problem for our present purpose, provided we look at diffusion and not viscosity. As a further simplification we shall first discuss the case of two dimensions, and later consider the extension to three.

In the leading order we have the familiar Boltzmann treatment in which only single collisions need be considered. It is part of the famous Stosszahl-Ansatz of Boltzmann that the molecule has no correlation with the target with which it is going to collide. If a molecule collides repeatedly with the same target, this process has to be described as a multiple collision rather than a succession of single collisions, since the absence of correlations would not apply to the later ones. The molecule cannot be re-scattered by the first scatterer unless it has in the meantime collided with another, so that we must consider triple collisions.

The simplest case is shown in Figure 5.1. The molecule collides with scatterer a, then with scatterer b, is scattered back by the latter, and collides again with a. For a given configuration of the scatterers, the number of such collisions into a final angular interval $d\theta$ is evidently, assuming the distance between the scatterers large compared to their dimensions,

$$I\sigma_a(\theta_1) \frac{1}{R} \sigma_b(\pi) \frac{1}{R} \sigma_a(\theta_2) \, d\theta, \qquad (5.1.1)$$

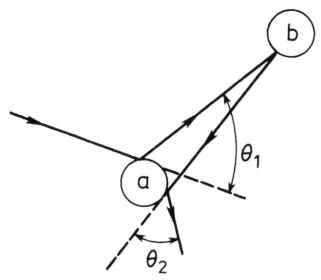

Figure 5.1   Triple scattering by two scatterers.

where I is the flux of incident molecules, σ_a and σ_b the differential cross sections per unit angle of the scatterers, and the angles θ_1 and θ_2 are as indicated in the figure.

This probability must now be integrated over the possible positions of the scatterers. The location of the first scatterer is arbitrary, and its position anywhere in the volume occurs with the probability ρ, the density of scatterers. If we assume the scatterers to be uncorrelated, which is a good approximation at large distances, in which we are interested, the probability of finding the second scatterer, b, at a position $\mathbf{R}$ relative to a is $\rho \, d^2\mathbf{R}$, which is $\rho R dR d\theta_1$. Apart from angular integrations, the integral of (5.1.1) therefore contains a radial integration, which for large R goes as $\int R dR / R^2$, and therefore diverges logarithmically.

It is now easy to see what will happen in the next order, without writing the expression in deatil. For the fourfold scattering shown in Figure 5.2, we require three angular densities which contribute factors of $1/R_{ab}$, $1/R_{bc}$ and $1/R_{ca}$ respectively. On the other hand

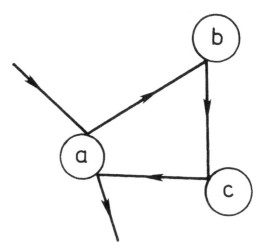

Figure 5.2   Quadruple scattering by three scatterers.

we have, after disposing of the first scatterer, elements of area $d^2\mathbf{R}_{ab}$ and $d^2\mathbf{R}_{bc}$. Keeping all angles and ratios fixed, this leaves us with an integral over the magnitude of the radii which goes as $\int R^3\, dR/R^3$, which diverges linearly. This is the term in the formal expansion which is proportional to ρ^3. Generally the term with ρ^n shows, for $n > 2$, a divergence of order $n - 2$.

In three dimensions the process corresponding to Figure 5.1 is finite, because the solid angle subtended by one scatterer at the position of the other is now proportional to $1/R^2$, so we have R^4 in the denominator and only $R^2\, dR$ in the numerator. However, the next term, of the type of Figure 5.2, has $R^5\, dR$ in the numerator and R^6 in the denominator, and thus is logarithmically divergent. After that the term proportional to ρ^n has a divergence of order $n - 3$.

Yet the physical problem of diffusion in a system of scatterers of finite density must have a finite solution. What is the origin of these divergences? The answer is that it is wrong to extend the integration over the various distances to infinity; the treatment has implied that no further scatterer intervenes, but this becomes very unlikely if the distance is greater than the mean free path.

A more realistic expression is obtained by inserting for each distance R traversed, a factor $\exp(-R/\ell)$, where ℓ is the mean free path. This factor gives the probability that no other collision has taken place over that distance. It is not claimed that this procedure is rigorous, but it gives a more realistic estimate of the processes we have specified than the naive expressions written before. Now everything is finite, and, for example in three dimensions, the nth term will be proportional to $\rho^n \ell^{n-3}$, if $n \geq 4$. But the mean free path is inversely proportional to the density of scatterers, so each of these terms is, in fact, proportional to ρ^3.

In other words, if we write down the terms involving multiple collisions in turn, we start, in three dimensions, with terms proportional to ρ and ρ^2, which are straightforward; but then we find a term in $\rho^3 \log \rho$, as well as an infinite series of terms proportional to ρ^3, which have to be summed if one is to get the result correct to third order in ρ. It is evident that this is far from easy, and that further terms will be harder still.

I am not qualified to review the progress that is being made in obtaining valid expressions with quantitative prescriptions for the coefficients. In presenting the results for the Lorentz model, I have followed the lectures by E. H. Hauge in "Transport Phenomena," *Sitges International School of Statistical Mechanics*, 1974, ed. G. Kirczenow and J. Marro, Springer, 1974. Hauge stresses the usefulness of this model which makes evident the divergences that in the case of mutual collisions between molecules were not noticed for some twenty years. He draws the moral: "Don't believe in a general scheme until it has been tested successfully on reasonable models!"

5.2. THERMAL CONDUCTION IN NON-METALS

Once it was understood that the thermal motion of non-metallic solids consisted of the vibrations of atoms around the equilibrium positions in the crystal lattice, and Debye had shown how, with the use of quantum theory, this accounted for the temperature dependence of the specific heat, it was also clear that the same lattice vibrations were responsible for the thermal conductivity in such bodies. In more modern terminology we say that the heat is carried by phonons.

Debye also was the first to attempt a quantitative description of this process. In summarizing his idea, it will be convenient to talk in terms of phonons, i.e., to use the language of quantum mechanics, although the argument is equally applicable at high temperatures, for which quantum effects are negligible, and a classical description would suffice. It turns out that the classical description would be more complicated, for a reason to which we shall return later.

In his theory of the specific heat, Debye had used the harmonic approximation, appropriate for small displacements from the ideal lattice, which leads to linear equations of motion. These equations would give an infinite conductivity, since phonons, according to this approximation, travel freely, and by setting up a distribution in which more phonons travel in one direction than in the opposite, one could have a steady flow of phonons carrying energy, without any temperature gradient. The situation is similar to that of a perfect gas, in which molecules do not collide. Such a gas also has an infinite thermal conductivity. While the collisions may be negligible for the specific heat, they are essential for causing the molecules to have a finite mean free path, and hence a finite heat conductivity. Debye pointed out that in a perfect crystal without impurities and imperfections, the free travel of phonons was limited by non-linear terms in the equation of motion, the so-called anharmonic force terms. The linear restoring forces for the atoms are only the leading terms in an expansion in powers of the displacement. Higher powers would, on dimensional grounds, be expected to be smaller in the ratio of the displacements to the interatomic spacing, and this ratio is small for most solids even near the melting point. Hence the harmonic theory of the specific heat is a very good approximation.

The next term, which is vital for the thermal conductivity, will make a quadratic contribution to the forces, i.e., will cause in the potential energy a term cubic in the displacements. A detailed computation of these cubic terms and their effects can be quite complicated, and Debye found an attractively simple way of avoiding the detailed study, as he did in so many other problems of physics.

He pointed out that the non-linearity in the equations of motion could be looked at as a density dependence of the velocity of sound, or of the refractive index for sound. The thermal motion of the lattice then causes fluctuations in the density, and hence in the refractive index, which will scatter sound waves. This model is rough in neglecting the dispersion, i.e., the dependence of the sound velocity on wavelength, and the effects of transverse lattice vibrations, which do not cause density changes. But these errors are similar to those in Debye's model of the specific heat, which gives an excellent qualitative description.

This way of looking at the problem makes it completely analogous to the scattering of light by density fluctuations, for which the theory was developed long ago. One only has to replace the refractive index for light by that for sound to obtain the scattering cross section, and hence the mean free path. The result was a prediction that the thermal conductivity should be inversely proportional to the temperature, which is in reasonable agreement with the data for high temperatures. Since the model is classical, it would not be expected to apply at temperatures below the Debye characteristic temperature, Θ.

The surprise in this case is that in spite of its appealing plausibility, Debye's way of looking at the problem is quite wrong, and its reasonable agreement with experiment fortuitous. The reason for this lies in the conservation of pseudomomentum, which was mentioned in section 4.2.

The effect of the cubic terms in the potential, which can be regarded as a small perturbation, can be studied by considering the matrix elements between unperturbed states. These are combinations of free-phonon states. Since the atomic displacements are linear in the amplitudes of the lattice vibrations, by a relation of the form of (4.1.2), the terms containing the third powers of the displacements are also cubic in the phonon amplitudes. The matrix element of each q_k links states in which the phonon number differs by unity, so the cubic terms have matrix elements in which three phonon modes change their occupation number by one phonon each. The only important processes in which energy can be conserved are then those in which two modes, with wave vectors $\mathbf{k}_1$ and $\mathbf{k}_2$ lose a phonon each, and $\mathbf{k}_3$ gains one, or vice versa. (In three dimensions there are at least three phonon modes for each $\mathbf{k}$, but we suppress the additional labels, for simplicity.) So the transition of greatest interest may be expressed symbolically as

$$\mathbf{k}_1, \mathbf{k}_2 \rightarrow \mathbf{k}_3. \qquad (5.2.1)$$

From the conservation of pseudovector discussed in section 4.2, we know that the transition is possible only if

$$\mathbf{k}_1 + \mathbf{k}_2 - \mathbf{k}_3 = \mathbf{G}, \qquad (5.2.2)$$

where $\mathbf{G}$ is a vector in the reciprocal lattice, such as appears in Bragg's law for diffraction of waves by a perfect crystal. One possible choice of $\mathbf{G}$ is always zero.

In addition, the transition cannot be a real, as opposed to a virtual, transition, unless energy is conserved. Since the energy of a phonon is $\hbar\omega$, where ω is its frequency in radians/sec, this requires

$$\omega_1 + \omega_2 - \omega_3 = 0. \qquad (5.2.3)$$

The appearance of $\mathbf{G}$ on the right-hand side of (5.2.2) is due to the atomic structure of the crystal. For a continuous medium with a non-linear response to distortions we would have the same equations but with $\mathbf{G}$ always being zero. In that case the conservation laws are of a very similar form to the conservation of momentum and energy in collisions. We therefore then have a situation like that of a gas in an infinitely long pipe with smooth walls. Even if the collisions between its molecules are allowed for, the momentum along the pipe is conserved. If the total momentum is non-zero, it will remain so in spite of collisions, and a flow of gas, also carrying energy, is a stable situation, not requiring a temperature gradient or pressure gradient to drive it. The reason that in practice the thermal conductivity of a gas is finite is that in measurements one closes the ends of the pipe, so there can be no net flow of gas, and hence no non-zero total momentum. In the case of the crystal, phonons can be generated or absorbed at the ends, so a phonon flux is not ruled out.

The objection is sometimes raised that the sum of the wave vectors, or the total pseudomomentum, which is conserved, is not proportional to the energy flux, and it does not follow that non-zero pseudomomentum implies non-zero energy flux. The reply is that the phonon interactions by the processes satisfying the selection rules (5.2.2) and (5.2.3) must lead to statistical equilibrium subject to the constraint of total pseudomomentum. This state is easy to specify, and from its symmetry it is evident that it contains a non-zero energy flux. We conclude that a continuous medium without atomic structure cannot have a finite thermal conductivity. The argument evidently remains valid if we include processes of higher order, involving four or more phonon modes.

Yet Debye's argument appears valid for a continuous medium. What has gone wrong? The answer here is that the model accepted too readily the results derived for the scattering of light as valid for the present problem. The standard theory of light scattering by density fluctuations treats the fluctuations as static regions of a varied refractive index; in other words, the rate of change of the density due to the motion of the underlying lattice waves is neglected. In the case of light this is indeed quite negligible, since the density fluctuations can move only with the velocity of sound, which is some 10^{-5} times smaller than light velocity. But the neglect becomes highly suspect when we consider the scattering of sound waves, whose velocities are precisely those of the density waves causing the fluctuations. It is only by including the full dynamics that we can arrive at the conservation law (5.2.2) which is of fundamental importance.

In a real crystal the atomic structure does cause transitions with non-zero **G**, which are called Umklapp processes (from a German word for "flip-over"), since one is typically concerned with the interaction of two fairly short-wave modes traveling, say, to the right, resulting not in a mode going with an even greater wave vector to the right, but instead in one traveling to the left. Such Umklapp processes, whose presence is vital for a finite thermal resistance, become rare at low temperatures.

The range of possible values of **k** comprises what is known as the first Brillouin zone, formed by drawing all possible vectors **G** from the origin, and all the planes bisecting these vectors. The order of magnitude of the maximum values of **k** is the inverse lattice spacing apart from a factor 2π. For three of these vectors to add to **G**, at least one of them must be of a length greater than $\frac{1}{3}$**G**. This means that its energy must exceed some finite value, less than, but of the order of, the maximum of the acoustic spectrum, and the number of such phonons present at low temperatures is, by Planck's law, of the order of

$$\exp(-\hbar\omega/KT)$$

We therefore expect an exponential rise of the thermal conductivity at very low temperatures, and this prediction, made by the author in his thesis in 1929 (see Peierls, *Quantum Theory of Solids*, 1955, §2.4), was verified by Berman (*Proc. Roy. Soc.*, A208, 90) in 1951.

The last finding depends on quantum effects, but the restrictions (5.2.2) and (5.2.3) and the treatment of high-temperature conductivity do not, and it should be possible to express our arguments on these points in classical language. Yet this turns out to be surprisingly inconvenient. In that language our variables are the amplitudes of the various modes. It is then easy to derive the conditions (5.2.2) and (5.2.3) for the selection of modes coupled to each other. It is also easy to use energy conservation to show that, in the coupling corresponding to the example discussed, the gain in intensity by mode 3 equals the sum of the losses in intensity of modes 1 and 2. But we find one more relation is needed to complete the equations. It turns out that the missing condition is the one corresponding to the statement, in quantum language, that in the process the occupation numbers of each of the three modes change by one. Classically this says that, if ΔI is the change in intensity of a mode,

$$\frac{\Delta I_1}{\omega_1} = \frac{\Delta I_2}{\omega_2} = -\frac{\Delta I_3}{\omega_3}. \qquad (5.2.4)$$

This relation is, of course, true in classical mechanics as well as in quantum mechanics, and can be proved, but unless one is led by the quantum case to expect it to be valid, it can easily be missed.

The history of the problem of thermal conductivity by phonons contains several further surprises, and perhaps one of the most unexpected is that even today there are open problems. However, the rest of the story would take us too far into technical details.

5.3. PERTURBATION THEORY IN TRANSPORT PROBLEMS

Transport problems frequently involve a system with many degrees of freedom, which are nearly independent, but coupled by the interactions necessary to ensure approach to thermal equilibrium, and to make the transport coefficients finite. The discussion of thermal conductivity in non-metals in our preceding section is one example of this, with the part of the interaction played by the anharmonicity. Other examples are the electric and thermal conductivity of metals, where the electrons are normally regarded as moving nearly independently, but subject to interactions with phonons, and with impurities and imperfections.

In all these cases it is natural to regard the interaction as weak and treat it as a small perturbation. The way this arises is by formulating a Boltzmann equation, for example, for the number of electrons in a given state, by which the rate of change of this number due to the interaction is expressed as

$$\left(\frac{dn(\mathbf{k})}{dt}\right) = \sum_{\mathbf{k'}} \{B_{\mathbf{k}\mathbf{k'}} n(\mathbf{k'})[1 - n(\mathbf{k})] - B_{\mathbf{k'}\mathbf{k}} n(\mathbf{k})[1 - n(\mathbf{k'})]\}, \quad (5.3.1)$$

where the first term represents the number of electrons scattered into the state $\mathbf{k}$, and the second term those scattered out. The third factor in each term is due to the Pauli principle, representing the probability of the final state being empty. Labels other than wave vector, and the dependence of the coefficient $B_{\mathbf{k}\mathbf{k'}}$ on other degrees of freedom, e.g., phonon states, have again been suppressed.

If the interaction is weak, the transition probability B can be expressed in terms of Fermi's "Golden Rule":

$$B_{\mathbf{k},\mathbf{k'}} = \frac{2\pi}{\hbar} |\langle \mathbf{k}|W|\mathbf{k'}\rangle|^2 \delta(E_i - E_f), \qquad (5.3.2)$$

where $\langle \mathbf{k}|W|\mathbf{k'}\rangle$ stands for the matrix element of the interaction linking the electron states $\mathbf{k}$ and $\mathbf{k'}$, and also any relevant states of phonons or other degrees of freedom. E_i and E_f are the initial and final energies, including the electron energy and everything else that may change.

To appreciate the limit of validity of the Golden Rule, one must remember that it is derived from time-dependent perturbation theory, using the result that, if a system is initially in a state i, the probability of it being after a time t in the state f, is

$$|\langle i|W|f\rangle|^2 D(t), \qquad (5.3.3)$$

where $D(t)$ is defined by

$$D(t) = 2\frac{1 - \cos(E_i - E_f)t/\hbar}{(E_i - E_f)^2}. \qquad (5.3.4)$$

For any fixed t, D depends on the energy difference. This dependence is shown in Figure 5.3. It represents a peak of height $(t/\hbar)^2$, and of a

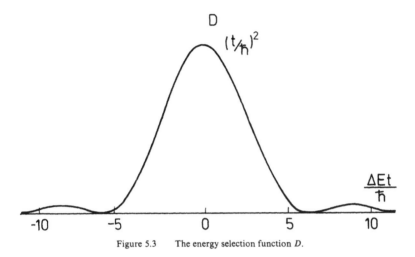

Figure 5.3   The energy selection function D.

width of the order $\pi\hbar/t$, so the area is proportional to $t/\hbar$, the actual value of the area under the curve being $2\pi t/\hbar$. For sufficiently large times, the width of the peak is negligible, and we may replace it in the limit by a δ-function. This leads to the Golden Rule, and for long enough times there is no question of its validity (subject always to the neglect of higher powers of the interaction W).

But in using the method to find the transition rate to be used in the Boltzmann equation (5.3.1) we cannot make the time t too long. If it is longer than the time τ between collisions, double and higher multiple collisions may have taken place, which are formally represented by terms of higher order in W, and these would then not be negligible. So t has to be less than τ. But this makes the width of the D function greater than $\hbar/\tau$. We may still neglect this width, provided the variation of other factors in the Boltzmann equation with energy is small over this width. Now the distribution function $n(k)$ certainly changes appreciably when the energy is changed by KT. It therefore looks as if the usual treatment depends on the condition

$$\hbar/\tau \ll KT. \qquad (5.3.5)$$

Failing this condition it would not seem to be right to rely on exact energy conservation in the transitions. This is not to imply that the law of conservation of energy may be violated, but this law applies to the total energy, including the interaction term. In a normal collision between isolated objects the interaction energy vanishes before and after the collision, and then energy conservation applies to the unperturbed energy, as expressed by the δ-function in the Golden Rule. But if the time between collisions is less than the duration of a collision, the interaction is never negligible, and conservation of the unperturbed energy never applicable. The implication of the condition (5.3.5) is then that we can be sure the collision is over, so completely that the interaction energy is less than KT, only after a time greater than $\hbar/KT$.

Now an order-of-magnitude check on a number of metals shows that the two quantities compared in (5.3.5) are generally of the same order of magnitude, sometimes one being larger, sometimes the other. In other words, the inequality is never well satisfied, and often strongly violated. This makes it look as if the electron theory of metals rested on very shaky foundations. This surprising suspicion of a trusted method was countered by a further surprising argument by Landau.

He saves the situation at least for those cases in which the interaction does not transfer energy to or from the electron. This applies evidently to scattering by impurities or lattice imperfections, since they are so much more massive than the electron that their recoil in a collision is negligible. It also applies to interactions with phonons at temperatures above the Debye temperature, Θ. In electron-phonon interaction the basic process is one in which a phonon is either created or destroyed. But $K\Theta$ is defined as the maximum energy of an acoustic phonon, so above Θ the phonon energy is less than KT, and an energy change by less than KT is statistically unimportant. Apart from quantum effects, the electrons see the disorder caused by lattice vibrations as static obstacles (as do light waves, according to what we noted in the last section).

In the case of a static potential, we may imagine having solved the Schrödinger equation at some energy E for an electron in the irregular potential. We would never in practice be able to write this solution down, but in principle it exists. It contains all the randomness which causes scattering, and in an external field therefore acquires a finite electric current. If the scattering is weak enough, this solution can be expanded in powers of the interaction, hence in powers of $1/\tau$. The limit up to which the expansion converges rapidly enough for the first term to be dominant must depend on τ, but the condition cannot be (5.3.5) because temperature has not yet been mentioned. Landau concludes that the condition must, in fact, be

$$\hbar/\tau < E. \tag{5.3.6}$$

For the metal, with its many electrons, we add the currents produced by the various electrons, and since the dominant contribution comes from electrons with energies near the Fermi energy, the net effect is to replace the condition (5.3.5) by

$$\hbar/\tau < E_F, \tag{5.3.7}$$

where E_F is the Fermi energy. This condition is well satisfied for all metals; there may be doubts about semiconductors and semimetals.

In cases in which the energy transfer is substantial we cannot apply Landau's argument, and therefore cannot show that the restrictive condition (5.3.5) is not necessary. However, there are few cases in which this could give trouble in practice.

Our description of the Boltzmann equation, and of Landau's resolution of the difficulty, are now old-fashioned, and recent presentations tend to use a much more sophisticated language. They nonetheless have to make an argument equivalent to Landau's, unless they are satisfied with deriving a series expansion without questioning its rapidity of convergence.

In this problem the net result is that the simple-minded approach which does not probe too deeply gives for most purposes the right answer, and that the doubts raised by a more thorough examination can ultimately be shown to be goundless. This is a not uncommon situation, which Pauli liked to call the "law of conservation of sloppiness".

6. MANY-BODY PROBLEMS

6.1. OFF-SHELL EFFECTS IN MULTIPLE SCATTERING

The standard problem of scattering theory is to determine the scattering amplitude when the interaction potential between the projectile and the target is given. There are, however, many cases in which one is interested in the inverse problem of finding the interaction potential from the observed scattering. This is the case, for example, in nuclear physics, where the mutual scattering of nucleons is an important source of information about the nucleon-nucleon interaction. In carrying out such an analysis one difficulty arises from the fact that scattering experiments give only the differential cross section, which is the square modulus of the amplitude, or, in the case of spin, the sum of squares of several amplitudes. The quantity of theoretical interest is the scattering amplitude.

For the present purpose we shall not discuss the methods by which one can get information about the amplitude from the cross section and other experiments (polarization, depolarization, etc.), but we shall assume that the scattering amplitude is known. How far does this determine the nature of the interaction?

If we may assume that the interaction is static, i.e., that there is an interaction potential depending only on the relative position of the two particles, but not on their relative momentum, there are well-known theorems showing that the potential is unique—in fact it is overdetermined. The point is that the scattering amplitude in any one partial wave as a function of energy is enough to determine the potential uniquely if there are no bound states, and that in the presence of n bound states the potential is determined by the knowledge of the amplitude at all energies, the energies of the bound states, and n further parameters. Knowing the scattering amplitude as a function of energy and angle gives us the knowledge of all partial waves, of which the higher ones usually have no bound states (the only bound state of the two-nucleon problem is an S state), so that any one of them in principle determines the potential.

The interaction may not be static, however. The usual models for the nucleon-nucleon force contain a spin-orbit term which depends on angular momentum, and a more complicated momentum dependence is not ruled out. If we do not restrict ourselves to static interactions, the scattering amplitude does not determine the potential uniquely, and there exist many so-called "phase-equivalent" potentials which give the same phase shifts, and hence the same scattering amplitude.

This ambiguity is often expressed by the statements that the observed scattering is that on the energy shell, and that we lack information about the "off-shell scattering." The origin of this terminology is the following. If the potential is known, the Schröinger equation determines the wave function for the relative motion in the whole of space. The behavior of this wave function at infinite distance is given by the superposition of incident and scattered waves, so it is directly related to the scattering amplitude. But the wave function at finite distance is not directly fixed by the asymptotic behavior.

Scattering theory is often formulated in momentum space, in which we are concerned with the Fourier transform of the wave function. For scattering at energy E, the transform of the wave function has contributions from all momenta, not only from those with the momentum of a free particle of energy E:

$$\frac{p^2}{2m} = E, \qquad (6.1.1)$$

where m is the reduced mass of the relative motion. The sphere in **p**-space defined by (6.1.1) is called the energy-shell, and all other values of **p** are off-shell.

At large distances from the target the interaction is negligible, so the wave function behaves like that of a free particle, and therefore has a wavelength corresponding to (6.1.1). The Fourier transform of the wave function

$$\phi(\mathbf{p}) = \frac{1}{(2\pi\hbar)^{3/2}} \int e^{-i(\mathbf{p}\cdot\mathbf{r})/\hbar} \psi(\mathbf{r})\, d^3\mathbf{r} \qquad (6.1.2)$$

therefore has contributions from large $\mathbf{r}$ when $\mathbf{p}$ is on-shell, and ϕ there has a δ-function type singularity. For off-shell $\mathbf{p}$ the distant parts of the integral cancel by interference and only a finite contribution from small $\mathbf{r}$ remains.

In order to complete our knowledge of the interaction, we must therefore supplement the information from two-body scattering by some further evidence on the off-shell part. A plausible suggestion for this is to observe multiple scattering, i.e., to observe successive scatterings of a projectile by a number of similar targets placed fairly close to each other so that the wave scattered by one has not approached its asymptotic form before reaching the next target. For this the distance between the targets should be comparable with the wavelength, or less, and comparable with the size of the target.

In such a case one would expect the results to be sensitive to the off-shell part of the scattering amplitude. A. M. Baqi Bég decided to test this in a prototype case of double scattering. To have a good test case he assumed two phase-equivalent potentials. One was static, the other a "separable potential", such that in the general expression for a non-static potential,

$$V\psi(\mathbf{r}) \equiv \int V(\mathbf{r}, \mathbf{r}')\psi(\mathbf{r}')\, d^3\mathbf{r}', \qquad (6.1.3)$$

the kernel is a product

$$V(\mathbf{r}, \mathbf{r}') = \lambda g(\mathbf{r})g(\mathbf{r}'). \qquad (6.1.4)$$

The static potential was chosen in such a way that it vanishes outside a sphere of radius a, and the phase equivalence then ensures that the function $g(\mathbf{r})$ also vanishes beyond radius a.

The calculation was restricted to s-wave scattering from both targets, which is dominant if a is much smaller than the wavelength. To everyone's surprise the result of the calculation was that the double scattering did not depend on which potential was chosen, provided the targets did not overlap, i.e., provided their distance was greater than their diameter $2a$.

This surprising result did not look like an accident, and suggested that there was some general reason for this identity. A little thought led indeed to the reason: Consider the first target and the wave incident on, and scattered from, it. From the knowledge of the on-shell scattering we know its asymptotic form at large distances. But we also know that it satisfies Schrödinger's equation in all space. Moreover, we know that the interaction is confined to distances less than a, so in the whole of space outside the sphere of radius a the wave equation is that of a free particle. This means it fixes the wave function uniquely from its asymptotic form.

In other words, we conclude that in the space outside the target the wave function describing the scattering is independent of the nature of the interaction, given only the on-shell scattering. This applies to the scattering by one target. For the double scattering we have to take into account the further modification of this wave by the presence of the second target.

But from the first step we know exactly the wave incident on the second target. We can represent it as a superposition of plane waves, and by assumption we know the scattering amplitude of each; superposition of these yields the total double-scattering amplitude.

The wave function we have now constructed does not saisfy the wave equation at the first target, and therefore has to be further modified, but this correction is already not part of double, but rather of triple scattering. We can repeat our reasoning for triple collisions (it would just be more long-winded to express it) and the results again are insensitive to the details of the scattering interaction.

It is therefore clear that the result is much more general than the particular model chosen for the calculation, and applies also to multiple scattering of any order, subject only to the essential conditions of no overlap between the targets.

The finding is instructive because if the scattering is analyzed in momentum space, as is customary, terms involving off-shell scattering make appreciable contributions, as originally expected, and these contributions are sensitive to the details of the interaction. The reason is that the Fourier transform (6.1.2) includes in the integration also the region $r < a$, where the wave function does depend on the nature of the interaction. Only when all the contributions are put together to construct the double-scattering amplitude will the sensitivity cancel out.

It is interesting to note that, as long as we discuss the problem only in momentum space, this cancellation remains a complete mystery, and it is easiest to see its physical origin by the coordinate-space argument which we have used above. One lesson of this instructive example is therefore that it is not necessarily always best to work in momentum space, and that the choice of variables should be subject to the nature of the physical situation to be studied.

The problem has some relevance to the question whether scattering of nucleons from nuclei could help us obtain information on the off-shell part of the nucleon-nucleon scattering, and therefore complement the information from nucleon-nucleon scattering experiments. The finding is not strictly applicable because nucleon-nucleon forces do not have a sharply limited range, and the interaction potential dies away gradually. These "tails" of the potentials due to different nucleons will always overlap to some extent. Furthermore, the nucleons inside a nucleus are not static targets, but are in motion, and have an appreciable probability of getting so close to one another that even the strong parts of their interaction potentials overlap. Bég's theorem thus does not assure us that the nucleon-nucleus scattering is identical for different phase-equivalent potentials between nucleons. It does, however, suggest that the sensitivity may be much less than one would have guessed, and that therefore this method of discriminating between potentials is very difficult.

6.2. PERTURBATION THEORY IN MANY-BODY SYSTEMS

Many important problems involve the presence of a large number of identical particles. As usual, one does not have much hope of an exact solution of the Schrödinger equation, and if the system in question resembles one for which an exact solution is known, it is tempting to use perturbation theory to construct one from the other. A careful theoretician will then always enquire about the limits of applicability of perturbation theory—we would like it not only to be convergent, but sufficiently rapidly convergent to be satisfied with the first few terms, which are usually all we can hope to calculate. It is here that one is liable to meet surprises.

To understand the special features of this problem we consider first an extremely simple model, namely, the ground state of N non-interacting particles, which may be identical bosons, or distinguishable particles but with the same dynamical properties. They move in a potential $U(\mathbf{r})$, which we want to approximate by a similar potential $U_0(\mathbf{r})$, so that

$$W(\mathbf{r}) = U(\mathbf{r}) - U_0(\mathbf{r}) \tag{6.2.1}$$

is a small perturbation. Let $u(\mathbf{r})$ and $u_0(\mathbf{r})$ be the ground-state eigenfunctions of one particle in the potential U or U_0, respectively. Their difference is assumed small and expandable in powers of W by the usual perturbation theory, applied to a one-body problem.

Then we also know the ground-state wave function of the N-body problem, which for the two cases is:

$$\psi = \prod_i u(\mathbf{r}_i); \quad \psi_0 = \prod_i u_0(\mathbf{r}_i). \tag{6.2.2}$$

To see how similar they are, we compute their overlap:

$$\langle\psi_0|\psi\rangle = \prod_i \int u_0(\mathbf{r}_i)u(\mathbf{r}_i)\, d^3\mathbf{r}_i = [\int u_0(\mathbf{r})u(\mathbf{r})\, d^3\mathbf{r}]^N. \quad (6.2.3)$$

If the one-body eigenfunctions are normalized, so are the N-body functions (6.2.2). The last integral in (6.2.3) is less than unity by Schwartz's inequality, but if the perturbation is small it may not be much less than unity. If we call it $e^{-\alpha}$, the last form of (6.2.3) becomes $\exp(-N\alpha)$. This shows that for any given α, i.e., for any given magnitude of perturbation on each particle, the overlap between the original and the perturbed eigenfunction decreases exponentially with N. If our system relates to a macroscopic object, with N some 10^{23}, this is obviously disastrous for any prospect of expanding the wave function in powers of the perturbation, though in principle the expansion is still likely to converge. If we are concerned with a nuclear problem, where particle numbers are in the tens, the situation is not quite so disastrous, but still serious.

Insofar as perturbation theory is usually derived from a formalism which requires the expansion of the wave function in powers of the perturbing potential, its application to the simple many-body problem which we have looked at does not look encouraging. However, many quantities of practical interest are obviously going to come out correctly. For example, the energy is, in this case, the sum of one-particle energies. Because of the identity of the particles they are all equal, so the total energy is N times the energy of one particle. But the perturbation series for the one-particle energy eigenvalue in this case is precisely the same as if the other particles did not exist, and its rate of convergence is not affected by the value of N. The same is evidently true of the expectation value of any one-body operator, or sum of one-body terms.

The expectation value of any two-body operator can, in this non-interacting model, be expressed in terms of one-body expectation values, so the same statement applies. In fact the rapid convergence of the perturbation series (assuming W reasonably small)

fails only when one is considering operators involving a large number of particles, of which the N-particle wave function is a particularly extreme case, since it contains information about the simul-
taneous probability of all N particles taking certain positions, or
momenta.

We may look at the situation also in terms of the rough-and-ready
rule that the ratios of successive orders of perturbation theory are
given by a quantity of the form $W/\Delta E$, where W is a measure of the
perturbing potential and ΔE a measure of the spacing of the unperturbed energy levels. In the N-body problem the density of
possible excited levels is very much greater than in the one-body
problem because of the many ways of sharing a given excitation
between many particles, and this suggests poor convergence for
large N. At the same time the existence of these virtual excited states
is not relevant if there is no interaction, so we should be able to
avoid its consequences if we ask the right questions.

All we have said so far is applicable only to the special model of
non-interacting bosons, or distinguishable particles, which is hardly
of practical interest. However, the study of this model has led us
to ask questions which make good sense even when there is interaction, and it is not difficult to guess the answers for that case.
We visualize the interaction as pairwise

$$V = \sum_{i \times j} V(\mathbf{r}_i, \mathbf{r}_j), \tag{6.2.4}$$

though three-body and higher interactions would not alter the
picture, as long as the number of particles appearing in the interaction term is finite and reasonably small. The original and the
perturbed Hamiltonian may differ in the interaction V as well as
in the potential field U, including the interesting case in which the
interaction V is weak enough to be treated entirely as a perturbation.

It is fairly obvious that we must again expect the overlap between
the unperturbed and the perturbed eigenfunctions to be extremely
small. This can no longer be verified by as simple an argument as
equation (6.2.3), but it is easy to look at the first few orders of

perturbation theory and to see that they contain increasing powers of N, as do the terms in the expansion of (6.2.3). More elaborate arguments from many-body theory bear out this guess.

Similarly we guess that expectation values of one and two-particle operators (which include the total energy) will show a rate of convergence, if expanded in a perturbation series, which does not get worse for large N. Again one can make this plausible by examining the first few orders of perturbation theory, and verifying that terms in higher powers of N cancel. Textbooks dealing with many-body theory show how to write the whole perturbation series in a form in which this cancellation to all orders is manifest. The formalism required for this is known as the "linked-cluster" expansion, and it is an extension of a method developed in classical statistical mechanics in order to justify the approximations used to describe a nearly perfect gas.

Returning for a moment to the simple model of non-interacting particles, we note that the precise form (6.2.2) of the eigenfunction applies only to the ground state. For distinguishable particles the eigenfunctions of excited states can be written in similar form, except that the different factors now contain different one-body eigenfunctions. For the case of bosons, however, we have to replace the single product by a symmetrized product of one-body functions. One consequence of the symmetrization is that the number of states in which particles are in different orbits is reduced compared to the ground state or other configurations in which all particles are in the same orbit. This, in turn, gives rise to the "Bose condensation" of a perfect Bose gas. However, it does not affect the discussion of perturbation theory for this system.

In the presence of interaction, the symmetrization can be of great importance because it tends to increase the probability of any two particles being close to each other, by constructive interference. This can make the interaction more effective than it would be otherwise. An interaction which for distinguishable particles would be weak enough to yield a rapidly convergent perturbation series for one and two-body operators, may in the boson system give poor or no convergence. In particular, it seems likely that perturbation theory will fail when the conditions for condensation are satisfied.

It would take us too far to become involved with all aspects of the many-boson system, but we can extend to it the lesson to be drawn from our elementary model, that one must distinguish the applicability of perturbation theory to one-body and two-body quantities from its applicability to the complete wave function.

In the case of fermions, the wave function of the ground state is, even in the absence of interaction, no longer a simple product, but a Slater determinant:

$$\psi = \frac{1}{\sqrt{N!}} \begin{vmatrix} u_1(\mathbf{r}_1) \, u_2(\mathbf{r}_1) \cdots u_N(\mathbf{r}_1) \\ u_1(\mathbf{r}_2) \, u_2(\mathbf{r}_2) \cdots u_N(\mathbf{r}_2) \\ \cdots \\ u_1(\mathbf{r}_N) \, u_2(\mathbf{r}_N) \cdots u_N(\mathbf{r}_N) \end{vmatrix} ;$$

$$\psi_0 = \frac{1}{\sqrt{N!}} \begin{vmatrix} u_{10}(\mathbf{r}_1) \, u_{20}(\mathbf{r}_1) \cdots u_{N0}(\mathbf{r}_1) \\ u_{10}(\mathbf{r}_2) \, u_{20}(\mathbf{r}_2) \cdots u_{N0}(\mathbf{r}_2) \\ \cdots \\ u_{10}(\mathbf{r}_N) \, u_{20}(\mathbf{r}_N) \cdots u_{N0}(\mathbf{r}_N) \end{vmatrix} . \qquad (6.2.5)$$

If the perturbation expansion of the u_n in terms of the u_{n0} is rapidly convergent, we can, as before, be sure that we can expand one and two-body quantities, but not necessarily the whole many-body eigenfunction.

However, in this case the presence of many particles may actually improve the situation, and may make the perturbation less effective. The reason is that we may imagine each of the exact one-body eigenfunctions u_n represented as linear combinations of the u_{n0}. If we add to any of the first N eigenfunctions any multiple of any of the other $N - 1$ functions, the determinant is not altered. Matrix elements of the perturbation W which link two occupied states thus have no effect on the eigenfunction of the ground state of the whole system, and we may therefore omit this part of the perturbing potential W.

Whether this results in a substantial weakening of the perturbation depends on its nature. A short-range, rapidly varying potential tends to have matrix elements between very distant states, and only a small part of it will connect low-lying one-particle states. A long-range, slowly varying potential tends to have matrix elements mostly between closely adjacent states, and will therefore be weakened considerably by the omission of matrix elements between occupied states.

In any event, the matrix elements which remain significant must connect an occupied state (i.e., one among the first N) with an empty one, and this is associated with an appreciable energy change, unless the occupied state is one of the highest. This reduces the effectiveness of these parts of the perturbing potential.

We conclude that, for non-interacting fermions, the perturbation expansion of one or two-body expectation values may actually converge more rapidly than for a single particle or a pair. Again the expansion of the whole many-body wave function is much worse. In estimating how much worse it is we must bear in mind that the perturbation of the one-body eigenfunctions is strongest for those in the highest states and less for the deeper ones. The extra factors appearing in the many-body expansion therefore should not be taken as of the order of powers of the total number N, but rather of the (smaller) number of particles in strongly affected states.

We again expect similar statements to hold for interacting particles, and again formal many-body theory bears this out. One way of formulating this is in terms of an interaction between the fermions from which the matrix elements between occupied states have been omitted; this leads to the so-called Bethe-Goldstone equation.

The basis of our discussion of this topic, including the surprises we noted, are largely contained in the thesis of J. Lascoux, University of Birmingham, 1959.

6.3. Positronium Formation in Metals

Early experiments on the annihilation of positrons in metals appeared to show that the rate of annihilation did not increase with the electron density from one metal to another, as one might have expected. In trying to explain this behavior, the suggestion arose that perhaps the positron might couple with an electron to form positronium, and the lifetime of this system, if it existed, might be independent of the electron density, or only weakly dependent.

This immediately raises the question whether a positron at rest inside a Fermi gas of electrons can bind an electron to form positronium, and whether the electron density at the position of the positron will show the influence of this. To throw light on the problem we shall once again start from a simplified model. Replace the positron by a heavy object, so that it may be assumed to be at rest, and its recoil due to interaction with the electrons may be neglected. Also neglect the mutual interaction of the electrons, except for the screening of the "positron." We shall also disregard the periodic potential of the lattice and take the electrons as free. We are then concerned with free electrons in the potential of a screened charge, which we shall call $V(\mathbf{r})$. An interesting physical quantity to look at is the electron density at the position of the charge, which we shall take as the origin.

There are now two possible ways of approaching this problem. The first, and most obvious one, is to solve the Schrödinger equation for one electron in the field of the screened charge, i.e., in the potential $V(\mathbf{r})$, and then build up the many-electron wave function from the one-electron solutions so obtained. The electron density at any point, in particular at the origin, is then a sum of the densities due to all these states.

For a charge of $+e$ and a realistic screening radius, there will be just one bound state. It is, however, convenient to keep the charge q as a parameter (but keep the screening radius fixed). For a small value of q there is then no bound state; at a certain value, say q_1, a bound state first appears, at a higher charge, say q_2, a second bound state starts, etc. All other electrons are in states in the continuous spectrum, so that the density at the origin is

$$\rho(0) = \rho_1(q, 0) + \rho_2(q, 0) + \cdots + \rho_c, \qquad (6.3.1)$$

where

$$\rho_1(q, 0) = \begin{cases} 0, q < q_1 \\ |\psi_1(0)|^2, q > q_1 \end{cases}, \quad \rho_2(q, 0) = \begin{cases} 0, q < q_2 \\ |\psi_2(0)|^2, q > q_2 \end{cases} \cdots, \quad (6.3.2)$$

and

$$\rho_C = \int_0^{k_F} dk |\psi(k, 0)|^2. \qquad (6.3.3)$$

Here $\psi_1, \psi_2, \ldots$ are the bound-state eigenfunctions, and $\psi(k, \mathbf{r})$ is the eigenfunction for an unbound state, normalized per unit wave number. We may think of the latter as expressed in polar coordinates. Then only s waves contribute to the integral since the wave function of other partial waves vanishes at the origin.

Considering the density now as a function of q, we expect it to be non-analytic at q_1. This is because ρ_1 is certainly non-analytic at that point, since it vanishes for q less than q_1, and is non-zero beyond it. The behavior there is actually as const.$(q - q_1)$, but we do not need to use this result. In any case, ρ_1 behaves qualitatively like the lower curve in Figure 6.1. The contribution from unbound electrons is also influenced by the potential, but it is likely to be less sensitive, and we would expect it to be regular at q_1. We guess, therefore, that the total density, which, up to q_2 is just the sum of ρ_C and ρ_1, will behave like the upper curve in the figure, i.e., it will have a kink at q_1 and therefore be non-analytic there. It then follows at once that the perturbation series, which is an expansion in powers of the potential, i.e., of q, cannot converge beyond q_1. We shall see, however, that our guess is not correct.

It is instructive to look at the problem in an alternative way. This uses the consideration, discussed in the previous item, that the matrix elements of the potential which connect occupied states have no effect on the many-body wave function, since their effect cancels out in the determinant wave function of the form (6.2.5). In solving the Schrödinger equation in a potential in which these

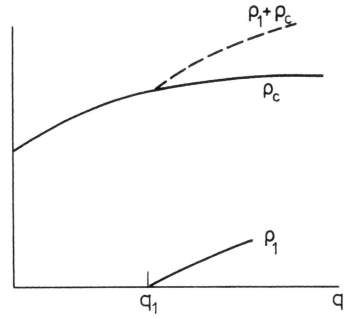

Figure 6.1   Conjectured electron density in a Fermi gas containing a screened positive charge. ρ_c contribution from the continuous spectrum, ρ_1 contribution from first bound state. Broken curve: total density.

matrix elements are included, we therefore only take the trouble of computing terms which are going to cancel out in the final answer.

This suggests starting from the Schrödinger equation in which the potential $V(\mathbf{r})$ is replaced by a truncated potential, which is defined like the Bethe-Goldstone potential, except that it is concerned with one particle only, and therefore is simpler. Its definition is

$$\tilde{V}\psi(\mathbf{r}) = V(\mathbf{r})\psi(\mathbf{r}) - \int_0^{k_F} dk \int d^3r\, u_0(\mathbf{r}, k)u_0(\mathbf{r}', k)V(\mathbf{r}')\psi(\mathbf{r}'), \quad (6.3.4)$$

where $u_0(\mathbf{r}, k)$ is the wave function of a free particle at wave number k normalized per unit wave vector; if we are interested only in the effect of the potential on s waves, we can restrict the eigenfunctions u to be s waves also. k_F is again the Fermi wave vector.

This truncated potential is non-local, since for given $\mathbf{r}$ in (6.3.4) the value of the wave function at other points $\mathbf{r}'$ enters. It would therefore in practice be inconvenient to work with.

We note that this potential is appreciably weaker than the original screened potential $V(\mathbf{r})$, and its first bound state will not occur until q is considerably larger than q_1, the charge for which the bound state appears in the one-electron problem with the potential $V(\mathbf{r})$. This suggests that perturbation theory with this truncated potential is convergent beyond q_1.

The evaluation of the perturbation series in terms of $\tilde{V}$ would look rather complicated. We note, however, that this series is, term for term, identical with the perturbation series for the original potential V. This is because perturbation theory preserves the anti-symmetry of the eigenfunctions, and is therefore at all stages compatible with the Pauli principle. Hence transitions to states which are already occupied must cancel, and the matrix elements belonging to them cannot appear in the answer.

This appears to show that the original, naive, perturbation expansion in terms of V should converge beyond q_1, contradicting the view we came to by considering the first method. How are we to resolve the contradiction between two conflicting, though admittidely not rigorous arguments? The weakest step is the guess, in the first argument, that ρ_C is regular at q_1. There would be no contradiction if ρ_C also had a kink at q_1, of such a sign and magnitude as to cancel that due to ρ_1. This situation is illustrated in Figure 6.2.

This is indeed what is happening. Consider the purely outgoing and purely incoming solutions of the radial Schrödinger equation for the s wave, $f(k, r)$ and $f(-k, r)$, which behave asymptotically as $\exp ikr$ and $\exp(-ikr)$. In terms of these the actual wave function, normalized per unit k, is

$$r\psi(k, r) = u(k, r) = \frac{1}{2\pi i \sqrt{2}} \left[e^{i\delta} f(k, r) - e^{-i\delta} f(-k, r) \right], \quad (6.3.5)$$

where δ is the usual s-wave phase shift. For ψ to be regular at the

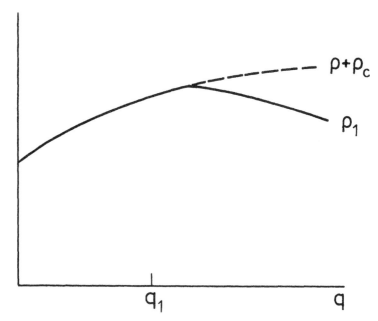

Figure 6.2   As Figure 6.1, but showing the actual behavior.

origin, u must vanish there, and this gives a relation between the phase shift and the "Jost function" $f(\pm k, 0)$:

$$\frac{f(k, 0)}{f(-k, 0)} = e^{-2i\delta}. \tag{6.3.6}$$

The density at the origin involves the value of $\psi(0)$, which is also (du/dr), taken at the origin. By using (6.3.5), (6.3.6), and the Wronskian identity

$$f'(k, r)f(-k, r) - f(k, r)f'(-k, r) = 2ik \tag{6.3.7}$$

one easily verifies that

$$\left[\left(\frac{du}{dr}\right)^2\right]_{r=0} = \frac{k^2}{2\pi^2 f(k, 0)f(-k, 0)}. \tag{6.3.8}$$

We may then write the integral (6.3.3), exploiting the symmetry in k:

$$\rho_C(0) = \frac{1}{4\pi^2} \int_{-k_F}^{k_F} \frac{k^2 \, dk}{f(k, 0)f(-k, 0)}. \tag{6.3.9}$$

In this form it is easy to discuss the nature of the dependence of $\rho_C(0)$ on q. It is well known that $f(k, r)$, defined as solution to the Schrödinger equation with well-defined boundary conditions, depends analytically on the parameters in the equation, including q. Hence the denominator of the integrand is analytic in q, and trouble can arise only from the zeros of $f(k, 0)$ or $f(-k, 0)$.

The Jost function $f(k, 0)$ cannot have any zeros for real k, except possibly at $k = 0$. There are zeros for complex k. In particular, a bound state has a wave function with the asymptotic behavior

$$u(r) = \text{const. } e^{-\kappa r} \tag{6.3.10}$$

which, apart from the constant factor, is just the definition of $f(k, r)$ with

$$k = i\kappa. \tag{6.3.11}$$

The bound-state radial wave function is therefore proportional to $f(i\kappa, r)$, and since it must vanish at the origin, the Jost function has a zero at $i\kappa$. Zeros in the lower half-plane of k correspond to resonances.

For q less than q_1 there is no bound state, and therefore no zero in the upper half-plane. Beyond q_1 there is such a zero, and since the Jost function is regular, the zero cannot appear from nowhere; it must at q_1 cross the real axis from negative to positive values.

Looking at (6.3.9) we see therefore that as q rises through the value q_1, a pole crosses the real k axis upwards, while another (that due to $f(-k, 0)$) crosses downwards. At this point, therefore, the argument for the analyticity of ρ_C fails.

We can now construct a function which is analytic in the neighborhood of q_1 by changing the contour of integration as q varies, so as to evade trouble with the poles. This means for any q above q_1 keeping the contour above the pole which was originally below the real axis, and below the one which was originally above. This requires a contour of the form shown in Figure 6.3, where the crosses indicate the poles, and the arrows their direction of motion with increasing q.

The integral along this contour equals the integral along the real k axis, which is ρ_C, plus the residues of the integrand at the poles. It follows that the sum of ρ_C and the two residues is analytic. The residue at the upper pole is

$$\frac{1}{2\pi} \frac{k^2}{\dfrac{df(k,0)}{dk}} f(-k,0) \qquad (6.3.12)$$

taken at $k = i\kappa$. The residue at the other pole is opposite and equal, but since the contour passes it in the opposite sense, it contributes the same amount as (6.3.12). It turns out that twice (6.3.12) is just equal to the density of the bound state at the origin.

Indeed, since the bound-state radial eigenfunction is proportional to $f(-i\kappa, r)$, from (6.3.10), we have, with proper nromalization:

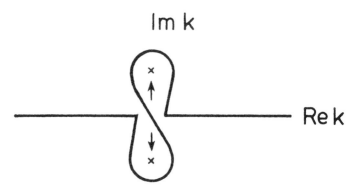

Figure 6.3   Complex poles in the integral (6.3.9) and shape of the contour avoiding crossing.

$$\rho_1 = \frac{[f'(i\kappa, 0)]^2}{4\pi \int_0^\infty [f(i\kappa, r)]^2 \, dr}. \tag{6.3.13}$$

On the other hand one easily derives from the wave equation, by considering two infinitesimally different values of k in the upper half-plane so that $f(k, r)$ vanishes for infinite r, the identity

$$f'(k, 0) \frac{df(k, 0)}{dk} - f(k, 0) \frac{df'(k, 0)}{dk} = 2k \int_0^\infty [f(k, r)]^2 \, dr,$$

and, in particular for $k = i\kappa$, since $f(i\kappa, 0) = 0$,

$$f'(i\kappa, 0) \left(\frac{df(k, 0)}{dk}\right)_{k=i\kappa} = 2i\kappa \int_0^\infty [f(i\kappa, r)]^2 \, dr \tag{6.3.14}$$

and also from the Wronskian equation (6.3.7), since $f(k, 0) = 0$:

$$f'(k, 0) = \frac{2ik}{f(-k, 0)}. \tag{6.3.15}$$

The last two equations establish the equality of ρ_1 (6.3.13) with the resultant of the two residues. It follows therefore that the sum of ρ_C and ρ_1, which by (6.3.1) is the total density just above q_1, is an analytic function of q. We see that the curves must indeed behave as sketched in Figure 6.2.

One conclusion to be drawn from this surprising result is that the question whether there is, or is not, a bound state for an electron at the bottom of the Fermi distribution has no unique answer, unless we specify clearly the description we wish to use. In our first method we had to include the bound state among the one-electron states we treated as occupied. In the second method we are justified in using perturbation theory starting from free electrons, and there is no mention of any bound state.

Returning now to the positron problem, we can obviously expect perturbation theory still to be applicable. The fact that the positron is as light as the electron has the effect of weakening the attraction, so that the binding energy of one electron to the positron is less than it would be to a heavy positive charge. The same is true for the effect of the electron-electron interaction, which we have so far ignored. Both the finite positron mass and the mutual interaction of the electrons exclude, of course, the use of our first method, and the only practical approach is by perturbation theory. The treatment of the electron-electron interaction has to be somewhat sophisticated, since part of it is already included in the screening which has been assumed, and one must avoid double counting.

The surprising conclusions discussed here were first reached in the 1959 Birmingham thesis by D. Butler, and published in the *Proc. Phys. Soc.*, 80, 741, 1962. In this paper the analyticity of the total density is shown for a square-well potential, but the general argument given above appears to be new. Butler's paper also examines the zero-order and first-order terms in the perturbation series for the square well, and shows them to be a good approximation to the exact answer for a sufficiently large k_F. The situation is even more favorable for an exponential potential. These results then allow him to apply the perturbation method with confidence to the real positron problem.

7. NUCLEAR PHYSICS

7.1. THE SHELL MODEL

An important tool in discussing the spectra and other properties of nuclei is the shell model, in which one starts from the approximation of regarding each nucleon as moving independently in a potential well, which is supposed to represent the mean effect of all the other nucleons. This has to be supplemented by finding the right combinations of those states which, in the absence of the inter-particle forces, would be degenerate with each other. This model was used from the very beginning of nuclear theory, long before there was any possibility of either an empirical verification, or a theoretical analysis of its justification. In fact, early versions of the shell model were used before the discovery of the neutron, when nuclei were still believed to contain protons, electrons, and α particles. I propose to review the way in which our understanding of the basis of the model has developed, which was by no means simple or straightforward.

It is probable that in the beginning the main motivation for using the shell model was habit. It had been the common method for dealing with atomic problems, and there it had been eminently successful. In atomic physics the concept of the gradual filling of shells by adding more and more electrons underlies our understanding of the Periodic System of elements.

It is also by far the simplest and most manageable approach to a many-body problem, and it was therefore very natural that, in exploring the new field of nuclear dynamics, one should have tried to use the shell model. It was also natural that one should use for the theoretical treatment of the shell model the same formalism as in atomic calculations, which was the Hartree-Fock method.

In this case one uses as the wave function of the nucleus—at least in the case of a closed-shell configuration—a Slater determinant of single-particle wave function. These single-particle functions are the eigenfunctions of a nucleon moving in a potential, the "Hartree-Fock potential", which represents the mean potential acting on each nucleon due to the others. The simplest prescription for defining this is to consider a Slater determinant with arbitrary single-particle functions, and calculate the expectation value of the real Hamiltonian (kinetic energy plus two-body interactions) for this function. One then requires the single-particle functions to be chosen in such a way as to minimize this expectation. This leads to functions satisfying the one-body Schrödinger equation with the Hartree-Fock potential, which is defined in terms of the one-particle functions, so that an actual calculation has to determine the potential and the one-particle functions together.

The use of a Slater determinant neglects all correlations between the nucleons, except those required by the Pauli principle. Two identical nucleons (both neutrons or both protons) with parallel spin must not be at the same point in space, and are unlikely to be close to each other; this feature is correctly represented in any antisymmetric function. However, two unlike nucleons or two like nucleons of opposite spin have an increased, or decreased, probability of being close together if the interaction between them is attractive, or repulsive. This effect is neglected, since the effect of the forces of interaction on the wave function is neglected. In the atom, where the interaction between any two electrons is rather weak, and slowly varying, such "dynamical" correlations are small, and their neglect reasonable.

In the nucleus the two-body forces are strong, and their range much less than the diameter of the nucleus, as is evident from the fact that nuclear forces saturate. This means that as the mass number A increases, the volume of the nucleus increases proportionally with A, and the binding energy per particle is approximately constant, suggesting that each nucleon interacts strongly only with a limited number of neighbors, so that it has to derive its whole binding energy from this interaction with very few other nucleons.

As a result, many physicists felt very strong doubts about the applicability of the shell model. These doubts appeared to be reinforced by the success, in the mid-nineteen-thirties, of Niels Bohr's explanation of the occurrence of many sharp resonances in the capture of slow neutrons by nuclei. Bohr's picture involved strong coupling between many nucleons, so that the additional energy contributed by adding a further neutron to the nucleus was almost immediately shared between many degrees of freedom. This makes the re-escape of the neutron difficult, because of the low probability of all the excess energy again being collected on one particle, which would allow it to escape from the attractive field of force. The lifetime of such a state can therefore be very long, and its energy, by the uncertainty relation, correspondingly sharp.

Bohr liked to picture the nucleus as a kind of liquid drop, and this view seemed to be the antithesis of the shell model. Nobody would expect a water molecule in a drop of water to move in its own way in some general field of force, without specific correlation with other molecules.

Another, apparently quantitative, argument was provided by a calculation first done by Euler (*Z. Physik*, 105, 553, 1937). He used the Hartree-Fock method to calculate the energy of nuclear matter, i.e., for the limiting case of an infinitely large nucleus. For this purpose one has to ignore the electrostatic repulsion between the protons, since the energy would otherwise tend to $+\infty$ in the limit, for any given density. While this nuclear-matter energy is somewhat academic, it is yet an interesting quantity since it can be obtained from the known energies of nuclei by extrapolation, correcting for the electrostatic contribution.

The Hartree-Fock calculation is very simple for this limiting case, since in an infinite volume there must be complete translational invariance, and therefore the one-particle eigenfunctions must be plane waves. The calculation thus consists only in evaluating the assumed two-nucleon interaction over a determinant of plane waves. There was no precise knowledge of the nuclear forces, but at the time they were believed to be reasonably well known. They had to give the binding energy of the deuteron right, which fixes, approximately, the product of the strength of the attractive potential times the square of its range. For given deuteron energy the triton and α particle become more strongly bound for shorter range, so their energies give an estimate of the range of the force.

Euler chose a Gaussian form for the variation of the interaction potential with distance. The calculation was later repeated by Huby (*Proc. Phys. Soc.*, A62, 62, 1949) with a Yukawa law of force, with very similar results. The curves of Figure 7.1 are based on Huby's numbers, which are easier to extract from the published information than Euler's. The Hartree-Fock result is the solid curve, in which the ordinate is the energy per nucleon; the abscissa is a measure of the specific volume—the radius per nucleon in the formula

$$R = r_0 A^{1/3} \qquad (7.1.1)$$

where R is the total radius and A the number of nucleons. The cross indicates the empirical value of E/A and r_0 by extrapolation. We see that the Hartree-Fock result has its minimum at more or less the right radius, but the binding energy comes out hopelessly wrong.

Evidently this failure indicates either that the interactions assumed were wrong, or that the Hartree-Fock approximation fails. To test this, Euler, and later Huby, evaluated the next order in perturbation theory. If the difference between the actual two-body interaction and the Hartree-Fock potential, in which the Slater determinant solution would be exact, is regarded as a small quantity, one can obtain a perturbation series in powers of this quantity. There is no first-order correction (due to the fact that the set of wave functions used is chosen to minimize the Hamiltonian among all single particle functions) and corrections appear only in second order. The second order consists in fact in allowing approximately for the correlations which had been left out in the Slater determinant.

The result is shown in the dashed curve. We see that it lies below the full curve, thus giving an increased binding energy, as expected. However, the increase is far from sufficient to bridge the gap to the empirical value. We also see that the binding energy is increased by a substantial factor, and this was interpreted by Euler as showing that the perturbation was not small, and that the perturbation series was not rapidly convergent.

I had, at that time, another argument which convinced me that the approximation must be inadequate: It was known that the forces used by Euler or Huby could give approximately the right binding energy for the α particle. This was found by a variational calculation, which represented a lower limit on the binding energy resulting from these forces. One would expect, therefore, that for nuclear matter these forces would give at least the binding energy obtainable by combining sets of four nucleons into α particles. This energy is shown as the dash-dotted horizontal line in the figure. A result which does not come close to this line must therefore be due to an inadequate approximation.

In retrospect we know today that all these arguments were wrong, and that the Hartree-Fock approximation would have been quite acceptable for the forces then assumed. Starting from the most physical argument, that of Bohr, one must distinguish between the ground state of a nucleus and a highly excited state, such as the compound nucleus formed by adding a neutron to a nucleus, so that the neutron binding energy of 6-8 MeV is available. Near the ground state the density of energy levels is small, because few of the nucleons can accept small amounts of energy without ending up in a state already occupied by a similar nucleon, thus violating the Pauli principle. The two-body interaction therefore is not capable of modifying the wave function and causing correlations. The reduction in the effective strength of the interaction by the Pauli principle has already been discussed in items 6.2 and 6.3. In the context of nuclear physics it was stressed by Weisskopf (*Science*, 113, 101, 1951).

We know (cf. (3.1.1)) that the condition for validity of perturbation theory is that $W \ll \Delta E$, where W is a measure of the perturbation and ΔE a measure of the spacing of the energy levels. In terms of this criterion we note that the spacing of the unperturbed levels

near the ground state for a closed-shell nucleus is of the order of
MeV, and it is not unreasonable to expect a perturbation to be
less than this. By the time we have reached an excitation energy
of 6 MeV, the distances between levels are of the order of 100 eV
or so, and in this situation no ordinary perturbation expansion
can work; we are really in a strong-coupling situation, in line with
Bohr's picture.

As regards Euler's interpretation of the curves in Figure 7.1, it
must be remembered that the binding energy of a nucleus is a rather
small difference between a large potential energy and a slightly
smaller kinetic energy. According to the model used, the total

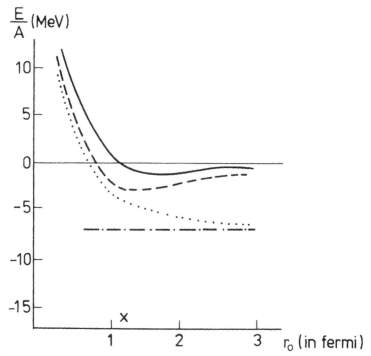

Figure 7.1   Energy of nuclear matter vs. specific volume, assuming "well-behaved" nuclear
forces, according to Huby. Full line: Hartree-Fock approximation. Broken line: including second-
order correction. Dash-dotted line: energy of free α particles. Dotted line: conjectured correct
answer. The cross indicates the energy and specific volume of nuclear matter extrapolated from
real nuclei.

potential energy might be of the order of -50 MeV per particle. The calculation is really a way of determining the potential energy, since the kinetic energy is not very sensitive to the details of the wave function. In order to see whether a correction is large or small, one should compare it not with the binding energy but with the potential energy. The difference indicated in the figure is then quite small, and we may conjecture that higher orders in the expansions would give even smaller contributions.

My argument fails because one can be sure of getting the α formation energy only at low density, when there is enough space for the α particles to exist without disturbing each other. This applies to the extreme right-hand end of the diagram. Here indeed the approximation fails, as my reasoning says it should. But at such low density the effect of the Pauli principle in weakening the interaction is small; if the interaction is weak enough to be treated in perturbation theory, it can only be at high density, where the Pauli principle helps. Of course in real nuclei we know that α particles attract each other, so that the lowest energy of a system of many α particles is less than that of the isolated ones, but we have no assurance that the "old-fashioned" forces used by Euler and Huby would lead to a mutual attraction of α particles. If they led to a mutual repulsion, the energy would rise on compression, and the "correct" result for the forces then assumed might look more or less like the dotted curve in the figure—reasonably approximated by Hartree-Fock for high density, but not for low density, and of course very different from reality.

But all this is hindsight, and for many years most theoretical nuclear physicists regarded the use of the shell model as an aberration. A few, out of stubbornness, or out of depth of insight, insisted on continuing with shell-model calculations. In particular, Maria Mayer felt convinced that the nuclear data gave evidence for the existence of shells in a similar way to the evidence for electron shells in atoms which appears in the Periodic System of elements. This evidence became too strong to be ignored. It remained difficult to make the neutron and proton numbers in the various shells correspond with the level sequence for any potential well. When Maria Mayer, and also independently Jensen, Haxel, and Suess, showed that a potential well with a spin-orbit coupling could give just the right numbers, the shell model became respectable and was universally accepted.

This implied also that the inter-nucleon forces were different from what had been assumed. This was borne out by high-energy nucleon-nucleon scattering experiments, which proved that the forces were less strongly saturating than had been thought (which left room for an attraction between α particles) and that they had a strong repulsive part at short distances.

The discovery of the repulsive core caused a new surprise because this repulsion prevents nucleons from approaching each other closely, and this amounts to a strong correlation. The Hartree-Fock method, which neglects all dynamic correlations, cannot possibly be applicable to a system containing these forces. Yet the shell model, which was regarded as synonymous with the Hartree-Fock approach, had proved itself.

This paradox was resolved only by the approach initiated by Brueckner, and developed by Bethe and others, which is based on the fact that the short-range interaction, in particular the repulsive core, acts in a way which is not very sensitive to the environment. We may therefore treat the interaction between two nucleons as taking place in the average field of the others, rather than assume each nucleon to move in an average field. It would take us too far to look at the Brueckner-Bethe method beyond this brief remark, or to review their quantitative findings, which are still undergoing revision. But they do make it possible to understand why the shell model can be applicable to forces for which the Hartree-Fock approach is not.

The surprise is, above all, how long it has taken, and how many detours have been gone through, before we have reached a reasonable general understanding of the situation.

7.2. CENTER-OF-MASS MOTION

One aspect of the dynamics of the nucleus which is very badly represented in the shell model approximation is the behavior of its center of mass. We know, of course, that in reality the momentum of the nucleus is conserved, and in consequence the velocity of its center of mass is constant. The shell model, on the other hand, implies the fiction that there is some field of force acting on all nucleons, attracting them towards its center, which is conveniently chosen as origin, so that the center of mass of the nucleons will oscillate about the origin. This oscillation is quite spurious, though it is an unavoidable byproduct of the shell-model approximation.

In this respect, therefore, the shell model wave function contains an error, which must be corrected if one is to replace the approximate wave function by an accurate one. How far is perturbation theory capable of correcting this error?

In discussing this question, we shall avoid the complexities of the Brueckner-Bethe formalism for taking care of strong short-range correlations by assuming that we are dealing with nucleon-nucleon interactions without strong repulsive cores. We are then entitled to base our shell-model approach on the Hartree-Fock method, and to correct it by ordinary perturbation theory. The lessons learned from this simplified model will then apply, in a more sophisticated way, to a more realistic, but more involved, treatment.

Our starting point is then again a Slater determinant, based on single-particle eigenfunctions in a potential well $U(\mathbf{r})$. The Hartree-Fock potential is actually a non-local potential well, expressed in terms of an integral operator, but all our subsequent arguments remain valid for that case. We shall not write down a non-local expression, for brevity.

The difference between the true potential energy and that used in our unperturbed wave function is then

$$W = V - U; \quad V = \sum_{i<j} V(\mathbf{r}_i, \mathbf{r}_j), \quad U = \sum_i U(\mathbf{r}_i). \quad (7.2.1)$$

The two-body potential V may also be non-local, and may depend on spin and isospin; we have suppressed these variables again, for brevity. Since we know that the shell model gives a reasonable representation of nuclear spectra, we would hope that the perturbation expansion in powers of W should converge. The correct application of perturbation theory involves, of course, the correct handling of degenerate states, but we may choose as example a closed-shell nucleus, with a non-degenerate ground state.

It then comes as a surprise to realize that in the circumstances envisaged the perturbation series does *not* converge. Moreover, this statement can be verified very easily, without performing any detailed calculations. To see this, it is convenient to choose as variables in place of the $3A$ coordinates $\mathbf{r}_i$, with $i = 1, 2, \ldots, A$, the center-of-mass position

$$\mathbf{R} = \frac{1}{A} \sum_i \mathbf{r}_i, \tag{7.2.2}$$

and the relative positions

$$\xi_i = \mathbf{r}_i - \mathbf{R}. \tag{7.2.3}$$

The ξ are not independent, since their sum vanishes identically by virtue of the definition (7.2.2), and we should therefore choose $A - 1$ of them as independent vectors, or an equal number of independent combinations. For our present purpose it is immaterial how this is done.

We then know that the exact solutions of the many-body Schrödinger equations can be written in the form

$$\Psi = e^{i\mathbf{k} \cdot \mathbf{R}} \chi(\xi). \tag{7.2.4}$$

Here $\mathbf{k}$ is the total momentum in units of $\hbar$, and ξ is a symbol for all the independent ξ_i. If the perturbation series converges, it should give an expression for (7.2.4) starting with the shell-model function which we shall call Φ. We know that perturbation theory always leads from ground state to ground state, so we expect that the expansion based on Φ for the ground state will lead to Ψ with $\mathbf{k} = 0$, which has the lowest translation energy.

In comparing Ψ with Φ, we may fix attention on the dependence on $\mathbf{R}$, for fixed values of all ξ_i. This is sketched in Figure 7.2, which shows Φ and Ψ as functions of a component of $\mathbf{R}$. Evidently Φ must be peaked, since the probability of the center of mass being far from the origin is very small. On the other hand Ψ is independent of $\mathbf{R}$. This comparison makes it clear that we are going from a discrete state to the edge of a continuous spectrum, and this is evidently impossible by perturbation theory.

The trouble can be further illustrated if we consider the special case in which the potential well underlying the shell-model function Φ is chosen to be harmonic:

$$U(\mathbf{r}) = \tfrac{1}{2}\alpha r^2. \tag{7.2.5}$$

$\sum_i U(\mathbf{r}_i)$ can be represented as a sum of two terms, one dependent only on $\mathbf{R}$, the other only on the internal variables ξ_i. Indeed, from (7.2.2) and (7.2.3) in this case

$$\sum_i U(\mathbf{r}_i) = \tfrac{1}{2}\alpha A R^2 + \tfrac{1}{2}\alpha \sum_i \xi_i^2. \tag{7.2.6}$$

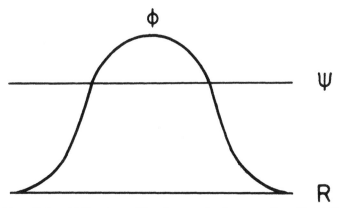

Figure 7.2   Shell model (Φ) and exact (Ψ) nuclear wave functions vs. position of the center of mass, for fixed values of the internal variables.

The ground-state wave function for this shell model is then also a product of an internal wave function and a Gaussian function of **R**.

In this particular case we have a better approach than the general perturbation expansion indicated above. We can subtract the center-of-mass term from the shell-model Hamiltonian (7.2.6), which leaves us with a potential acting only on the internal coordinates. The eigenfunctions of this problem can easily be constructed from those of the harmonic oscillator potential (7.2.5). We have now a purely internal potential and its difference from the true two-body interaction is also a function of the ξ_i only. In this modified form our objection to the perturbation expansion disappears.

However, this device is available only for the harmonic potential well, and the Hartree-Fock potential is not of that form. We can therefore use the harmonic potential (7.2.5) as an example for the direct application of perturbation theory.

If we define our perturbation W in the original form (7.2.1), both the unperturbed and the exact Hamiltonian separate into parts depending on **R** and parts acting on the ξ. The perturbation series then amounts to a combination of the perturbation of the expansion for the internal motion with that for the center-of-mass motion. Our difficulty concerns the latter. Let us write the ground-state eigenfunction for an oscillator with force constant $\alpha A(1 - \lambda)$. Then $\lambda = 0$ gives us the unperturbed problem, i.e., the **R**-dependent potential of (7.2.6), and $\lambda = 1$ the real case without center-of-mass force.

The ground-state eigenfunction of this problem, for mass $M = Am$, is

$$\psi(R) = \frac{[\alpha A^2 m(1 - \lambda)]^{3/8}}{(\pi\hbar)^{3/4}} \exp\left\{-\frac{1}{2\hbar}[\alpha A^2 m(1 - \lambda)]^{1/2}R^2\right\}. \quad (7.2.7)$$

It is now evident that this function has a branch point for $\lambda = 1$, i.e., that 1 is the convergence radius for the power series in λ. The series may just be conditionally convergent for $\lambda = 1$, but it is clearly of no practical use.

Evidently the same behavior will occur for the general shell model, which does not allow a separation of the form (7.2.6). It comes as a shock to realize that the perturbation expansion for correcting the shell model, used very widely at least for qualitative discussions, is open to such a fundamental objection.

For practical purposes it should be remembered, however, that one rarely gets beyond the second-order term in such a perturbation expansion. This will contain a contribution to the elimination of the zero-point energy of the center-of-mass oscillation wrongly included in the shell model. This contribution is in the right direction, if numerically not justified. The error thus caused to the calculated energy levels is no doubt less than the higher terms in the perturbation series for the internal motion, which are omitted anyway. The "invalid" expansion is therefore likely to give quite a fair indication of the nuclear states and their energies. The error becomes important only if we ask questions relating specifically to the center-of-mass motion, when the perturbation series becomes quite useless.

There are techniques available for getting around the difficulty. One is to add also to the exact potential V a fictitious center-of-mass potential, which can easily be eliminated at the end of the calculation. (See, for example, Sanderson, Elliott, Mavromatis, and Singh, *Nuclear Physics*, A219, 190, 1974). In that case one is applying perturbation theory to go from a discrete state to another discrete one, and there is no longer the certainty that the expansion cannot possibly be convergent. Whether it is in fact convergent, and whether it converges rapidly enough to be practically useful, needs a separate investigation.

Alternatively, one can use a projection method to eliminate the center-of-mass motion from the shell-model states, before applying perturbation theory. Since the projected states no longer form a complete orthonormal system of functions, one has to use a modified form of the theory, which has been developed by the author (Peierls, *Proc. Roy. Soc.*, A333, 157, 1973; see also Atalay, Mann, and Peierls, *Proc. Roy. Soc.*, A335, 251, 1973) but is not as yet much used in practice.

8. RELATIVITY

8.1. Radiation in Hyperbolic Motion

Our next, and last, surprise arises from an old paradox. Consider a charged body suspended, at rest, in a static gravitational field. Evidently it is surrounded by an electrostatic field, and there is no question of any radiation being emitted. By the principle of equivalence, which is the main basis of general relativity, this situation is equivalent to the object being in uniform acceleration in a flat space. We are accustomed to thinking that an accelerated charged body emits radiation. Yet the presence or absence of radiation should be an observable fact, and therefore not depend on the frame of reference. Unless one of the steps in our argument is wrong, we are facing a violation of the principle of equivalence.

We shall refer to the first frame of reference, in which there is a gravitational field, as the G frame, and the second, in which space is flat, as the F frame. In the F frame the object is accelerating indefinitely, and therefore its velocity cannot remain small compared to light velocity; it is therefore essential to use relativistic kinematics. However, we shall begin by using non-relativistic results to gain some general understanding, and later extend our discussion to the relativistic case.

The non-relativistic expression for the rate at which a small charged object emits radiant energy is usually written in the form

$$\frac{dW}{dt} = \frac{e^2}{6\pi\varepsilon_0 c^3} \dot{\mathbf{u}}^2 \equiv A\dot{\mathbf{u}}^2, \qquad (8.1.1)$$

where ε is the charge, and $\dot{\mathbf{u}}$ the acceleration. It is known that this expression (or its relativistic generalization, where appropriate) gives the right answer for the total energy loss by radiation in all practical situations. However, it is also evident that it is not the right expression for the instantaneous rate of loss of energy by the charged body.

We see this by considering the expression for the work done by the body:

$$\frac{dW}{dt} = -\mathbf{F} \cdot \mathbf{u},$$ (8.1.2)

where $\mathbf{F}$ is the force acting on the body, in this case the radiative reaction. If (8.1.1) were correct at every instant, we could, at least for rectilinear motion, obtain F by dividing by u, and this would suggest an infinitely large reaction force at the instant when the velocity, but not the acceleration, vanishes. Instead, we can integrate (8.1.1) by parts, to obtain

$$\frac{dW}{dt} = -A\mathbf{u}\ddot{\mathbf{u}} + A\frac{d}{dt}(\mathbf{u}\dot{\mathbf{u}}).$$ (8.1.3)

In integrating over a time interval, the second term contributes the difference in the values of $A\mathbf{u}\dot{\mathbf{u}}$ at the initial and final time. This difference vanishes in any periodic motion, if the interval consists of a number of complete cycles. It also vanishes in the case of a collision, such as in the emission of X rays, if the charged particle is initially and finally free, so that $\dot{\mathbf{u}}$ vanishes at the ends of the interval.

This is typical for almost all practical problems in which one is interested in the emission of radiation, so that we retain the agreement with the predictions for the energy loss if we replace (8.1.1) by

$$\frac{dW}{dt} = -A\mathbf{u}\ddot{\mathbf{u}}.$$ (8.1.4)

Now we have a perfectly reasonable expression for the radiative reaction force, in agreement with (8.1.2), namely,

$$\mathbf{F} = A\ddot{\mathbf{u}}.$$ (8.1.5)

Note that in the problem which gives rise to our paradox the overall effect of the second term in (8.1.3) is not negligible, since for uniform acceleration neither the velocity nor the acceleration vanishes before or after. Indeed the new form for the energy loss is by no means equivalent to the old one: (8.1.4) and (8.1.5) both vanish if $\dot{u}$ is constant, and we are tempted to conclude that there is no radiation in frame F, and this would restore the validity of the principle of equivalence. This resolution of the paradox would involve the surprising finding that uniform acceleration did not cause the emission of radiation.

This discussion can be extended to the case of relativistic kinematics. The easiest way of doing this is to note that the expression (8.1.5) for the radiative reaction force is valid at time t in the Lorentz frame in which the charged object is at rest at that time. By applying a Lorentz transformation to the equation we can find the force for an arbitrary velocity. We shall do so only for the case in which velocity and acceleration are in the same direction. The result is then

$$\mathbf{F} = A\left\{\frac{\ddot{u}}{[1 - (u/c)^2]^2} + \frac{3u\dot{u}^2}{c^2[1 - (u/c)^2]^3}\right\}. \qquad (8.1.6)$$

It is easily verified that, with the usual formula for the radiation loss,

$$\frac{dW}{dt} = A\frac{\dot{u}^2}{[1 - (u/c)^2]^2}, \qquad (8.1.7)$$

we have

$$\frac{dW}{dt} = -Fu + \frac{d}{dt}\frac{Au\dot{u}}{[1 - (u/c)^2]^2}, \qquad (8.1.8)$$

where again the last term does not contribute to the time integral if u or $\dot{u}$ vanishes at each end of the time interval, or the motion is periodic.

It is also easy to verify that the reaction force (8.1.6) vanishes for the "hyperbolic motion", in which the intrinsic acceleration is constant:

$$z = \frac{c}{g} \sqrt{(c^2 + g^2 t^2)}. \qquad (8.1.9)$$

The idea that this hyperbolic motion of a charged particle does not lead to the emission of radiation was first put forward by Pauli (*Theory of Relativity*, Pergamon Press, 1958, §32γ). He reached this conclusion from a consideration of the electromagnetic field, but it would certainly be consistent with the absence of a radiative reaction force.

But there are further surprises to come. A careful discussion by Fulton and Rohrlich (*Ann. of Phys.*, 9, 499, 1960) proved that there is radiation emitted. We must be careful in defining what is meant by radiation, since after long times the particle gets very close to the radiation it has emitted. Fulton and Rohrlich calculate the retarded potential, and hence the fields, caused by one particular point on the particle's world line, after a long time t, so that the fields are considered on a sphere of radius ct. One then determines the Poynting vector on this sphere, and hence the flux of energy through the sphere. Finally one takes the limit of infinite t. The result of this calculation is a rate of emission of radiation in agreement with (8.1.7).

This brings us back to the paradox about the equivalence principle, and raises a new paradox, because on the face of it the presence of emitted radiation and the absence of a radiative reaction force seem incompatible with conservation of energy.

Of these the second is resolved, following Fulton and Rohrlich, by realizing that the energy balance involves the energy radiated, the mechanical energy of the particle, and the "self-energy," i.e., the energy of the field surrounding the particle. In the usual case in which the particle is moving uniformly both before and after the emission of radiation, the field around the particle eventually settles down again to that of a uniformly moving charge, and its energy depends only on its velocity. However, in the hyperbolic motion, in which the particle keeps accelerating, the field is more complicated. In particular, it does not follow that it depends only on the velocity, since the fields are always those derived from retarded potentials, and this destroys the symmetry between the decelerating and the accelerating phase of the motion. Thus the field energy at the time when the particle is going out with velocity u need not be the same as that when it came in with velocity $-u$, and the difference is capable of making up for the energy of the radiation emitted between these times.

This leaves the paradox with the equivalence principle, and the way to resolve this is shown in some work, as yet unpublished, by D. Boulware. To make his point we have to specify the G frame explicitly. We require that in this frame the particle be at rest, and that the metric be time independent. (If the gravitational field in the neighborhood of the particle were changing in time, we would not be surprised to see radiation emerging even from a stationary charge.)

These requirements determine the G frame uniquely, apart from trivial changes of coordinates. If we call the coordinates in this frame ξ, η, ζ, τ, the relation with those in the F frame is

$$x = \xi$$

$$y = \eta$$

$$z = \zeta \cosh \tau$$

$$ct = \zeta \sinh \tau \qquad (8.1.10)$$

A particle following the hyperbolic motion (8.1.9) in the F frame is now stationary at $\zeta = c/g$. The line element is given by

$$ds^2 = c^2\, dt^2 - dx^2 - dy^2 - dz^2 = \zeta^2\, d\tau^2 - d\xi^2 - d\eta^2 - d\zeta^2$$

$$(8.1.11)$$

which, as required, is independent of τ.

Figure 8.1 shows lines of constant ζ and of constant τ in the z, t plane. The point to observe is that the part of space-time consisting of all positive ζ and all τ from $-\infty$ to $+\infty$ corresponds to the sector of the z, t plane between $ct = z$ and $ct = -z$. Similarly negative ζ are represented by the opposite sector, but with negative z. The rest of the z, t plane is not reached for any real ζ and τ. The physical reason for this relation is that in the G frame the gravitational field also deflects light, and with the metric (8.1.11) causes a focusing of the light rays, and therefore a horizon.

It now turns out that the radiation which Fulton and Rohrlich showed to be emitted all goes into the part of the z, t plane with $ct > z$, $t > 0$, which has no equivalent in the G frame. If this is

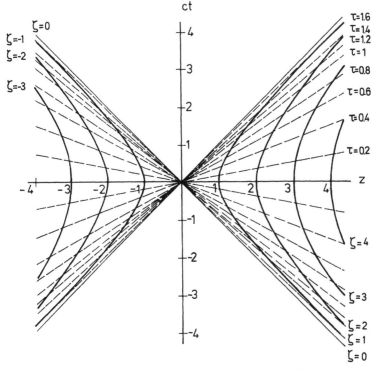

Figure 8.1   Lines of constant space coordinate and of constant time in the G frame, shown in the F frame.

correct, we have no violation of the equivalence principle locally, because that part of space-time which can be described in both frames carries no radiation on either description. There is radiation in that part of F space-time which has no equivalent in the G frame, and for it the question of equivalence therefore does not arise.

The proof that the emitted radiation does indeed go into that part of space-time is very simple. Consider the light signal originating at time t_0, when the charged object is at z_0. When this signal has traveled a distance R in a direction forming an angle θ with the z axis, it is at

$$z = z_0 + R \cos \theta, \text{ at time } t = t_0 + \frac{R}{c}. \qquad (8.1.12)$$

Therefore

$$ct - z = ct_0 - z_0 + R(1 - \cos \theta). \qquad (8.1.13)$$

The second term is non-negative, the first part negative. However, we are instructed to take the limit $R \to \infty$, for fixed z_0, t_0. In this limit the second term dominates, except for an angular region near $\theta = 0$, which becomes infinitely narrow in the limit. We can conclude that the amount of radiation escaping to infinity in the region describable in the G frame vanishes.

It might be added that the paradox is somewhat academic, since in reality gravitational fields never extend to infinity, and if the field we have described is embedded in a flat space, the complete equivalence, and with it the paradox, disappears. This is the argument used by Fulton and Rohrlich to avoid the paradox. There is nothing in the laws of general relativity or in electrodynamics, however, to exclude the extreme situation we have discussed, and it is therefore satisfactory that even in this idealized case the paradox can be resolved.